Cybercrime During the SARS-CoV-2 Pandemic (2019-2022)

Cybersecurity Set

coordinated by
Daniel Ventre

Volume 3

Cybercrime During the SARS-CoV-2 Pandemic (2019-2022)

Evolutions, Adaptations, Consequences

Edited by

Daniel Ventre
Hugo Loiseau

WILEY

First published 2023 in Great Britain and the United States by ISTE Ltd and John Wiley & Sons, Inc.

Apart from any fair dealing for the purposes of research or private study, or criticism or review, as permitted under the Copyright, Designs and Patents Act 1988, this publication may only be reproduced, stored or transmitted, in any form or by any means, with the prior permission in writing of the publishers, or in the case of reprographic reproduction in accordance with the terms and licenses issued by the CLA. Enquiries concerning reproduction outside these terms should be sent to the publishers at the undermentioned address:

ISTE Ltd
27-37 St George's Road
London SW19 4EU
UK

www.iste.co.uk

John Wiley & Sons, Inc.
111 River Street
Hoboken, NJ 07030
USA

www.wiley.com

© ISTE Ltd 2023

The rights of Daniel Ventre and Hugo Loiseau to be identified as the authors of this work have been asserted by them in accordance with the Copyright, Designs and Patents Act 1988.

Any opinions, findings, and conclusions or recommendations expressed in this material are those of the author(s), contributor(s) or editor(s) and do not necessarily reflect the views of ISTE Group.

Library of Congress Control Number: 2022950071

British Library Cataloguing-in-Publication Data
A CIP record for this book is available from the British Library
ISBN 978-1-78630-801-6

Contents

Introduction ix
Daniel VENTRE and Hugo LOISEAU

**Chapter 1. The Evolution of Cybercrime During the
Covid-19 Crisis** 1
Daniel VENTRE

    1.1. Introduction. 1
    1.2. Observing the evolution of cybercrime 4
        1.2.1. Leveraging annual data: the case of India 8
        1.2.2. Leveraging monthly data 11
        1.2.3. Leveraging weekly data: the case of China. 21
    1.3. Has the global geography of cyberattacks changed?. 29
    1.4. Conclusion 34
    1.5. Appendix 39
        1.5.1. Cybercrime tools: malware 39
        1.5.2. CVSS as indicators of vulnerability levels 40
        1.5.3. Heterogeneity and complexity of cybercrime typologies. 41
        1.5.4. Attitude of companies toward cyber risks: the case of
        the United Kingdom 46
    1.6. References 47

**Chapter 2. The SARS-CoV-2 Pandemic Crisis and the Evolution
of Cybercrime in the United States and Canada** 49
Hugo LOISEAU

    2.1. Introduction. 49
    2.2. The impacts of the SARS-CoV-2 pandemic. 50

2.3. Cybercrime and SARS-CoV-2 52
  2.3.1. Targets and victims............................ 53
  2.3.2. Malicious actors.............................. 57
  2.3.3. Cyberspace: a propitious environment for cybercrime 58
2.4. The evolution of cybercrime in North America during the pandemic. . 61
  2.4.1. The United States............................. 62
  2.4.2. Canada 67
2.5. Discussion 69
2.6. Conclusion 72
2.7. Acknowledgments................................ 74
2.8. References 74

Chapter 3. Online Radicalization as Cybercrime: American Militancy During Covid-19 81
Joseph FITSANAKIS and Alexa MCMICHAEL

3.1. Introduction.................................... 81
3.2. A new typology of cybercrime 83
3.3. Internet connectivity and violent militancy 85
3.4. The pre-pandemic domestic threat landscape 87
3.5. The domestic threat landscape of the pandemic 88
3.6. Pandemic accelerationism 91
3.7. From virtual to real-life criminality..................... 93
3.8. Online radicalization during Covid-19................... 94
3.9. A new methodological paradigm for online radicalization? 98
3.10. Conclusion: meta-radicalization as cybercrime 100
3.11. References.................................... 102

Chapter 4. Cybercrime in Brazil After the Covid-19 Global Crisis: An Assessment of the Policies Concerning International Cooperation for Investigations and Prosecutions................ 109
Alexandre VERONESE and Bruno CALABRICH

4.1. Introduction: Brazilian cybercrime and the Covid crisis impact 109
4.2. Cybercrime in the literature and the Brazilian case 112
4.3. A theoretical model for international cooperation 115
4.4. The evolution of cybercrime in Brazil 119
4.5. The evolution of the Brazilian legal system concerning cybercrime and its connection to the international regime 126
4.6. Managing international cooperation without having the best tools ... 133
4.7. Difficulties with cooperation: joints, mortises, and notches 137

4.8. Conclusion: what to expect from the future?	140
4.9. References	142
4.10. Appendix: List of interviews and questions	147

Chapter 5. Has Covid-19 Changed Fear and Victimization of Online Identity Theft in Portugal? 149
Inês GUEDES, Joana MARTINS, Samuel MOREIRA and Carla CARDOSO

5.1. Introduction	149
5.2. The impact of the Covid-19 pandemic on cybercrime	150
5.3. Evolution of cybercrime in Portugal	153
5.4. Online identity theft (OIT)	155
5.4.1. Definition and modus operandi	155
5.4.2. RAT applied to cyberspace	156
5.4.3. Individual variables and OIT victimization	159
5.5. Fear of (online) crime	160
5.5.1. Determinants of fear of (online) crime	160
5.6. The present study	162
5.6.1. Measures	163
5.6.2. Results	165
5.6.3. Variables associated with online victimization and fear of identity theft	169
5.7. Conclusion	170
5.8. References	171

Chapter 6. A South African Perspective on Cybercrime During the Pandemic 177
Brett VAN NIEKERK, Trishana RAMLUCKAN and Anna COLLARD

6.1. Introduction	177
6.1.1. Background to South Africa and the pandemic	178
6.1.2. Methodology	179
6.2. International rankings	180
6.3. Cybercrime and related legislation	183
6.4. Cybersecurity incidents	186
6.4.1. Ransomware	186
6.4.2. Scams and fraud	188
6.4.3. System intrusions and data breaches	190
6.4.4. Disinformation and malicious communications	192
6.4.5. Other	196
6.5. Discussion	197

6.6. Conclusion 199
6.7. References 199

List of Authors 211

Index 213

Introduction

Cyberspace is composed of several different layers that are essential to the functioning of an interconnected and functional network. The physical, software and informational layers, although forming the functional body, are of little interest to the political field. It is, on the contrary, the social stratum that concerns and interests politicians and political scientists in particular. This layer includes the set of individual behaviors interacting with cyberspace and a collective component that affects the policies, institutions, laws, norms, regulating and framing the interactions and use of cyberspace. Despite the fact that cybercriminals may be interested in flaws in software or physical systems to commit their crimes, the social character is also fraught with vulnerabilities that can be identified and exploited by cybercriminals. Considering that each user is responsible for his or her actions in the cyberspace and that cybersecurity practices are not always appropriate or sufficient, the user is simultaneously a potential victim and a system gateway. In that sense, systems are vulnerable from the very moment a user behaves unsafely.

By virtue of its supranational character, the cyberspace is particularly difficult to govern and secure. Based on the amount of information that passes through the networks at all times, supervising and controlling information and transactions, verifying content legitimacy and legality and processing complaints or incident reports within a reasonable amount of

---

Introduction written by Daniel VENTRE and Hugo LOISEAU.

time, constitute major challenges for supervisory, governmental, industrial, public or private bodies. The detection and recovery phases can be affected by those capacity limits, making systems non-operational, or exposing them to multiple risks.

Network security and protection are considered a shared responsibility between security and law enforcement agencies, government bodies, businesses, organizations and individuals. Several social changes have therefore taken place in recent years with the increasing dependence on and use of cyberspace. Policies and surveillance do not always keep pace with the advances in this space. This is especially true in the event of a crisis such as the one experienced by all the States during the SARS-CoV-2 (or Covid-19) pandemic in 2020.

During this health crisis, cyber risks and cyberthreats seem to have increased. Cyber risk is the product of the level of threat with the level of vulnerability. While cyber risk determines the likelihood of a successful cyber attack [SAN 22], cyberthreat represents a potential for violation of security, which exists when there is a circumstance, capability, action or event that could breach security and cause harm [SAN 22]. These two notions are fundamental, because the cyberization that societies have experienced over the past 20 years has contributed greatly to the complexity of cybersecurity issues. In 2001, U. Beck already heralded that the social production of wealth is accompanied by the social production of risks [BEC 92]. The SARS-CoV-2 pandemic in 2020 and its following multidimensional crisis (social, economic, political, etc.) represent a contextual window, which perfectly illustrates the risk society that Beck referred to, featuring the multiplication and diffusion of systemic and cross-sectoral risks resulting from technological and industrial developments. Globalization and the expansion of the cyberspace increase such risks [BEC 92]. During this period, cybercrime evolved in a context of globalization and cyberization.

Cybercrime research primarily focuses on two broad categories of crime:

– Organized cybercrime, large-scale cyber attacks, which could be part of the more general framework in the study of major crime. Jean-François Gayraud proposes four main characteristics for appraising "major crimes":

- i) major crime is chiefly manifest by polycriminality; major criminal groups are opportunistic and pragmatic in criminal markets, meaning that they do not necessarily develop a specialization of their criminal practices;

- ii) these groups are territorialized, rooted in a space that allows them to create their own biotope, setting up hermetic enclaves inaccessible to the public authorities; this favors the territorialized and immaterial expansion into the cyberspace;

- iii) these groups and organizations are considered unsinkable; this is because they are highly adaptable to socioeconomic changes and resistant to repression by the public authorities or to the competition from other criminal groups;

- iv) finally, these groups a have major macroeconomic impact since they manage massive, globalized and interconnected financial flows, which facilitate and stimulate corruption, as well as the laundering of illicit income [GAY 21].

– The scope of cybercrime and its study is not limited to organized crime, but includes everyday life, ordinary criminal acts, such as online defamation, violent extremism and hate speech on the Internet and social media [BEN 22], radicalization [ALA 21], disinformation [PAR 20], etc., which are all common forms of cybercrime.

Major criminal groups and ordinary cybercrime play a crucial role in what might be termed a "criminological diffuse background", whose omnipresence and pervasion in the cyberspace act as the basis for many illicit practices of interest to cybersecurity [BRE 10]. As a result, cybercrime benefits from an immense market bringing together the supply (of software, services and available techniques in the cyberspace) and the demand from criminal or non-criminal organizations, states and individuals whose goal is to exchange a good (data) by means of an increasingly dematerialized and fungible currency: the cryptocurrency [BAD 20]. While the aggregation of these characteristics has facilitated cybercrime, what can be said about the crisis context in 2020?

I.1. The context

I.1.1. *The pandemic, its management, its effects*

The SARS-CoV-2 epidemic emerged at the end of 2019, its first cases being recorded in China, and then in Thailand. China quickly implemented measures trying to contain the epidemic: massive population lockdown in

Wuhan, as well as the construction of field hospitals whose progress the whole world could follow live via webcams connected to the site on a permanent basis. Soon enough many cases and victims were identified in various parts of the world. The WHO officially announced the outbreak of this new disease on January 30, 2020 and declared it a global pandemic on March 11, 2020. As a result, several states in the world decided to implement emergency health safety policies. These policies were not implemented everywhere at the same time, or in the same way, given the fact that the epidemic did not evolve at the same pace around the world. In addition, governments often had conflicting approaches as to what should be done (to close the borders or not, to confine the whole population or only certain age brackets or professional categories, etc.). Measures took several forms:

1) Those aimed at fighting the epidemic itself:

– measures prohibiting or restricting movement within the states and/or internationally, the lockdown of national or local populations (districts, cities, regions), isolation of individuals and social distancing;

– measures restricting social, economic and professional activities: store closures, the reduction of international trading volumes and the closure of schools and businesses.

These measures were applied to varying degrees across the different countries – not in all states – at different times, and sometimes disparately even within the same country, following specific schedules for each region (as was the case in France, and is still the case in China in 2022, where the population of Shanghai, for example, has been forced into strict lockdown, while other regions have not been confined).

2) Some states also decided to implement measures to mitigate the negative effects produced by health constraints on society, in particular, the impact on the economy. It is worth recalling:

– economic security measures: state funding in order to help the economic activity of companies, business loans, emergency social benefits, etc.;

– measures to ensure the continuity of activities: not only remote working, distance learning, but also hybrid work formulas alternating teleworking and regular attendance to the office.

Due to its lethality and morbidity, and to the disruption of societies' working mechanisms, the Covid-19 pandemic is considered "a public health crisis without precedent in living memory [...] which brings with it the third and greatest economic, financial and social shock of the 21st century, after 9/11 and the Global Financial Crisis of 2008" [OEC 20]. According to the OECD report, the "shock" occurred at several levels:

– a halt or slowing down in production in countries affected by the pandemic and the lockdown phases;

– a disruption of supply chains across the world;

– a steep drop in consumption;

– a collapse in "confidence" (which is reflected in the fluctuations of financial markets confronted with a scenario of extraordinary uncertainty);

– the significant loss of human life;

– proof of the stark weaknesses of healthcare systems around the world, including in the richest countries.

Other elements could be added to this list, such as the emergence of conspiratorial and anti-vaccine movements.

This reveals the presence of *at least* two types of crisis whose effects are combined: a health crisis together with an economic and financial crisis.

I.1.2. *The concept of "crisis"*

The term "Covid-19" was quickly associated with that of "crisis".

Crises are particular moments of tension, during which certain phenomena or processes are exacerbated. The definitions of "crisis" highlight some of these characteristics: "the emphasis is on the idea of a sudden and intense manifestation of certain phenomena, signaling a rupture"; "a sudden and intense manifestation, whose duration (of a state or behavior) is limited, potentially leading to harmful consequences"; "a disturbing situation, due to a disruption of balance and whose outcome is decisive for the individual or society"; "a situation of deep unrest in which society or a social group is immersed, giving rise to fear or hope for a profound change"[1]. Briefly stated, a crisis is a temporary situation, a turning point, a moment of instability and stress. However, the aforementioned approach assumes the pre-existence of a state of normality – albeit temporarily interrupted – which must be regained. The return to the previous state will be considered to be the crisis resolution. Although the crisis represents a moment of exception, it is a condition one may still cope with.

For Fearn-Banks [FEA 09], a crisis is "a major occurrence with a potentially negative outcome affecting the organization, company, or industry, as well as its publics, products, services or good name. It interrupts normal business transactions and can sometimes threaten the existence of the organization" [FEA 09, p. 2].

But normality can also be considered to be the crisis: from that perspective, the history of the world would be a series of successive, superimposed, nested crises [CAR 18], constituting the normalcy of the world, its structural signature [KOS 72].

Crises are moments of tension, of disorganization having reached a threshold that an individual, a group of individuals, or a society can no longer find acceptable or tolerable. The peak of intensity cannot last. But it should also be noted that crises are maintained over time, which may seem incompatible with its very definition (economic, security, climatic, political, social crisis, etc.). Still, the notion of duration is highly subjective. Furthermore, it is perhaps not so much the duration that characterizes the crisis, but the level of disorganization or even destruction it engenders, as well as the acceptability threshold of such an exceptional state by an individual or society. The crisis signals a moment of unbearable tension. What is called the end of crisis, the after crisis, the exit or crisis resolution is a return to a level of acceptability, even if the disturbance persists. But it is simply less intense or perceived as such. What is called the crisis exit may not always refer to the end of the event itself, but denote the ability to coexist with it, or to partially accept it. Therefore, *risk tolerance* is an essential component in the definition of a crisis. As such, it may differ from one society to another. Tolerance acts as a threshold defining the presence of a crisis or its absence.

Finally, with the resolution or exit from the crisis, one cannot consider that it is possible to return to the normality prior to it because the crisis has

---

1. Available at: https://www.cnrtl.fr/definition/crise.

necessarily left a mark on society. One may observe there is a pre-crisis period (A), a crisis period (B) and a post-crisis period (C), where C differs from A, because C = A + the effects of B.

The notion may be manifest as an economic, financial, debt, social, demographic, systemic, political (a state crisis, a crisis of power, a crisis in confidence, a crisis of democracy, etc.), humanitarian, migratory, health, climate, environmental, national, international or global crime crisis [ALT 07]. Cybercrises refer to crises resulting from one or more cyberattacks. These are rare events with a strong impact, situations when one or more malicious action(s) on an information system cause(s) a major disruption of the entity, having various and significant impacts, and sometimes causing irreversible damage [ANS 21].

Crisis indicators have been designed to try to keep track of the many crises occurring worldwide. The "crises" identified[2] have their origin in armed conflicts, insurrection movements, civil wars, proxy wars, intra-state conflicts, conflicts for the control of territories and resources, leading to humanitarian (due to the scarcity of resources and massive displacement of populations), health, economic crises. Other indicators[3] propose a quantitative measurement of the degree of severity of humanitarian crises, in view of adapting the responses to be provided. The INFORM Severity Index, produced by ACAPS[4], identified 136 global crisis situations in 2022[5].

Crises are moments of destabilization experienced by societies, having a varying impact on social groups. Crime can exploit such moments to establish its influence, its presence and develop its activities. For example, Mexican drug cartels took advantage of the Covid-19 crisis to expand their influence into realms or areas of activity where the state was weakened or absent (protection theory) [JAS 19, KLE 14]: "With a health care system inaccessible to large portions of the population, as well as welfare programs put under extreme strain, criminal organizations have been observed

---

2. Available at: https://theowp.org/our-work/crisis-index/.
3. Available at: https://drmkc.jrc.ec.europa.eu/inform-index/Portals/0/InfoRM/GCSI/GCSI%20Beta%20Brochure%20Single.pdf.
4. ACAPS is a non-governmental initiative, supported by three NGOs: the Norwegian Refugee Council (NRC), Save the Children and Mercy Corps. Available at: https://data.humdata.org/organization/acaps.
5. Available at: https://www.acaps.org/sites/acaps/files/crisis/gcsi-download/2022-06/20220606_inform_severity_-_may_2022.xlsx.

distributing resources to some local communities. Though anti-drug efforts continue at the state and federal level, officials within the government have largely side-lined security in favor of prioritizing pandemic response"[6]. Facing the mask crisis [WAN 20], mask shortages and the inability to ensure sufficient industrial production, criminal actors rushed into the breach, hoping to take advantage of the expectations of populations. Criminal cyber-operations on the theme of Covid, masks and drugs have been carried out all over the world [EUR 20a].

The link between crime and crises has been the subject of numerous studies, particularly borrowing from the economic theory of crime [DEF 11], but also focusing on other categories of crisis such as wars, political crises and disasters. For example, Kontula [KON 97] considers that crime occurring in exceptional circumstances is marked by predatory behavior, the erosion of moral values and the reduction of fear of punishment, as well as by the loss of control of the situation by security actors. UNODC [UNO 12] argues that economic factors are important in the evolution of crime. But the analyses diverge: when addressing the effects of the 2008 international financial crisis, Kurtz [KUR 15] concluded there was no close relationship between the two phenomena. In Russia, on the contrary, periods of economic turbulence coincided with an upsurge in criminal activity in 1998, and later in 2008–2010 (economic crimes and crimes against property seem to have been highly reactive to changes in economic conditions) [IVA 12]. "Peaks" in criminal activity may occur during periods of economic crises or "economic stress".

I.1.3. *The role of cyber in crises*

Digital was quickly considered as one of the responses to certain aspects of the Covid-19 crisis: teleworking was supposed to partly guarantee the continuity of activities in several sectors; e-commerce was expected to support market activities; and the various online applications proposed by governments and/or private sector initiatives were to maintain the efficiency of the health system, organize the logistics of large-scale vaccination phases, manage the lockdown and traveling restrictions (through the use of tracing applications) and ensure the respect for social distancing (by checking the possession of a health pass to access the places requiring it). Furthermore,

---

6. Available at: https://theowp.org/crisis_index/mexican-drug-war-2/.

digital technology had to be a tool for creating social ties at a time when individuals were forbidden from the slightest face-to-face relationship, it had to take citizens out of their isolation and help them cope with situations which had only existed in works of fiction until then (the all-pervading images of deserted cities and millions of individuals at their windows, cloistered by force, awaiting liberation).

But while digital technology has contributed, in its own way, to the global effort to fight the spread of the virus, it has also exacerbated the already high degree of dependence of societies on communication technologies, in particular the Internet and cyberspace.

Did cybercrime feed on that particular context, taking advantage of such dependence, those vulnerabilities and the increase in the use of digital technology? As the above-mentioned indicators have shown, the crises in the world are multiple at a time t, simultaneous, distributed within the entire international system. These are all contexts in which crime, and consequently cybercrime, can evolve. This point will be essential in the present analysis: Covid-19 cannot be considered as a single, isolated crisis. It would be more accurate to refer to Covid-19 "crises", in the plural form, to refer to the crises resulting from the epidemic, its management and its effects on societies. These crises may be of a health, economic, social and perhaps political nature. But to these crises are added all those having existed before and during the Covid-19 epidemic, an event which took place in a world animated by tensions, conflicts and calamities and which is still immersed in discord. Crises coexist, become interlocked, can be triggered by the pandemic itself, or preceding it, for other motives, but may also become interdependent, producing effects on one another: the epidemic did not spare the populations already facing wars, other diseases, economic difficulties. In addition to the displacement of populations due to the climate crisis or to wars and economic or political crises, citizens had to cope with the effects of the pandemic itself. Cybercrime irrupted into societies hitherto affected by multiple crises to varying degrees. One cannot consider the evolution of cybercrime in the light of the crises strictly related to the Covid-19 pandemic, but should integrate them into a broader context made up of multiple crises, to which the pandemic was added.

I.2. Literature review: works on the theme "cybercrime and Covid"

I.2.1. *Main themes and hypotheses*

The effects of the epidemic on global society have inspired many reflections since the first months of 2020. These explore the impacts of the pandemic on:

– the economy: an increase in poverty together with a health crisis plunging millions of additional workers into it, as well as an increase in unemployment, with "around 205 million unemployed people in 2022, that is, a lot more than the 187 million in 2019" [ONU 21];

– culture [YU 21], education [ONY 20] and science [GUP 21];

– security and defense: the epidemic brought to light the weaknesses of the common European security and defense policy, highlighted the vulnerabilities of member states in terms of infrastructure, supply chain and communications security. The pandemic accentuated the retreat of the United States and the EU from the international scene, to the benefit of China, posing a challenge in several areas, including IT security and cyber capabilities. The pandemic has been described as an accelerator of pre-existing trends and an amplifier of instabilities [MEY 21]. The protection against the pandemic became a matter of national security: both the economic vitality of a nation and its way of life were endangered. Due to the spread of globalization, it influenced all states, whose destinies were more closely intertwined than they had been in the past centuries. The pandemic was destructive and disruptive. It disrupted or paralyzed the security and defense strategies of states, simultaneously exposed to several categories of threats or risks: crime, terrorism and foreign state threats. More generally, the Covid-19 pandemic was categorized as an event with a profound and lasting impact on the international security environment [ORO 22]. The Covid period favored and stimulated the development of new criminal activities [EUR 20b] (fraud, international trafficking, counterfeiting, etc.) feeding on global instability [KEN 21].

From the early months of 2020 – when the pandemic was still in its infancy – articles addressed the question of the evolution of cybercrime in such a context.

Some guiding themes and hypotheses have emerged from the abundant literature produced since then, both academic and non-academic (national and international organizations, cybersecurity companies, private and public sectors, etc.). The following should be retained:

– By reinforcing the essential role of the Internet, the management of the Covid-19 crisis (lockdown, teleworking, social distancing, tracking applications, e-commerce, etc.) created favorable conditions for cybercrime. Organized crime could take advantage of the enlargement of the attack surface by multiplying or diversifying criminal opportunities [TRI 20]. The conditions thus created acted both as a catalyst [BOU 21] and an accelerator of cybercrime.

– The multiplication of vulnerabilities and criminal opportunities acted as a key factor in the evolution of cybercrime since the pandemic onset: the lockdown transformed Internet uses, certain online practices such as e-commerce was auspicious for the theft of personal data; teleworking [TAB 20] isolated employees who could be the target of social engineering attacks [VEN 21]. The changes brought about by the pandemic in everyday life, particularly the uses of "cyber", played a central role in explaining the evolution of cybercrime.

– Attack vectors diversified, involving the creation of new attack scenarios [GRY 21]. The thriving of cybercrime during the pandemic was mainly the result of its ability to adapt, innovate and renew its operating methods [COR 20], its business model [LAA 21] and even the reconfiguration of some of its groups. ANSSI discusses the professionalization of organized cybercrime groups and the specialization process which has characterized the evolution of cybercrime in recent years [ANS 22]. Internet uses changed during the lockdown phases, shifting vulnerabilities or creating new ones: cybercrime also had to adapt to this reconfiguration of the attack surface in order to seize opportunities [LAZ 21]. For example, this was done via "themathized" operations (the registration of several tens of thousands of domain names using the term "Covid" and associated terms) [NAI 20], or by aiming its actions toward essential sectors in times of health crisis (health industry, vaccine research centers, hospitals, logistics, etc). The attacks on healthcare actors were at the core of the research carried out by Chigada and Madzinga [CHI 21]. While crime remained unchanged in its nature or composition, it simply adapted to the new scenario. Cybercrime polished its methods, its targets, in some cases even its organization, matching the new context in which it evolved.

Nevertheless, it should be noted that over the past 2 years, advancements within organized cybercrime have had other driving forces than the Covid crisis alone. One should bear in mind the necessary adaptation of operating methods, the attack tools used, the choice of targets, imposed by technical developments: in that sense, cybersecurity can make certain targets too resistant, require too much effort on the part of the attackers, making the intended targets less attractive. Certain skills may also become necessary while others are no longer required due to technical and technological advances.

– The scarcity of criminal opportunities in the confined "offline" world could have prompted a shift from "offline" crime to "online" crime [PLA 21]. This hypothesis has its detractors [MIR 21] for whom the shift does not occur from offline to online crime, but mainly within the online crime category.

I.2.2. *Theoretical frameworks*

Focusing on the transformations in the lifestyles and daily practices of hundreds of millions of individuals throughout that period at a global scale – something which offered new opportunities for crime – the routine activity theory [COH 79] became popular as the main explanatory framework [HAW 20, KEM 21, GOV 21, HOR 21, PLA 21, CHE 21, KOP 22, OLO 22, IMS 22].

According to this theory, a crime is likely to be committed when three conditions are met: the presence of a motivated offender, the presence of an accessible target to the offender/criminal and the absence of an efficient guardian. Target vulnerability increases when all three elements are present. Hawdon et al. [HAW 20] have argued that the societal changes forced by the compulsory lockdown quantitatively and qualitatively increased those conditions. Vulnerability, or in this case cyber risk, increased because the threat (the motivated malicious actor) and the vulnerability (the presence of suitable targets) converged on the same place (cyberspace) at the same time.

In Collier et al. [COL 20], low-level cybercrime (in terms of technical capabilities) may have increased due to the rise in the number of confined teenagers and young adults who seemed to be launching simple attacks against poorly protected networks just for fun and to earn a little money. This idea was further discussed by Payne [PAY 20] who also claimed that a significant victimization of people aged 50 and over (less equipped to digitally defend themselves and with poorer cyber-hygiene) was observed during the first lockdown wave. Cybercrimes such as pandemic-themed targeted frauds were particularly used by cybercriminals.

On the whole, there seems to have been both an increase in the number of delinquents motivated by the money and the recreational dimension and the presence of targets meeting the goals of cyberdelinquents, especially those lacking well-established security habits. To this should be added the fact that areas affected by the lockdown also included cybersecurity actors from private companies and government agencies. Unavoidable teleworking forced many employers to focus their attention on helping employees transfer work to their homes, but overlooking network securitization. As network security was not necessarily ensured by cybersecurity services, the responsibility of monitoring a larger part of the networks was entrusted to the police services. Regarding this point, Dupont explains that classic police methods prove insufficient in the case of imminent lockdown: "Classic police investigation and arrest methods are proving insufficient, as they are too slow to produce tangible results on a large scale. They are most effective when combined with innovative damage prevention and mitigation strategies" [DUP 20].

However, the impacts of forced lockdown were too rapid to deploy preventive measures. For example, mitigation advertising campaigns were launched toward the end of March and the beginning of April 2020 in Canada. The reduction in network protection capabilities and resources was a corollary effect of the first lockdown phase.

Several authors have based their analyses on the routine activity theory in order to account for the particular structure of opportunities having arisen as a result of lockdown regulations in the United States and the Canadian provinces. The increase in the attack surface, in the number of cybercrime actors with diverse motivations and resources, and network monitoring issues have all been mentioned as factors influencing user and network vulnerability. Added to this is the great adaptive flexibility of criminal groups, as described by Gayraud, which has facilitated the activity adaptation of such groups to the global pandemic context, by relying on the "diffuse criminological background" of cyberspace.

I.3. Our research questions

This work studies the role of cybercrime in the world and its evolution, during the early pandemic period (in the first days of the year 2020), which is still not over at the time of writing these lines. The case studies discussed contribute to reflections on the link between cybercrime and crises as well as on the explanatory variables of cybercrime.

I.3.1. *Chapter 1 – The evolution of cybercrime during the Covid-19 crisis*[7]

The dominant narrative since the first months of the Covid-19 epidemic expressed that, through the effects produced on societies, there was a significant (sometimes even spectacular) increase in cybercriminal activity. Not only the health crisis scenario but also its economic consequences were believed to foster a context favoring cybercriminal activity, as well as increasing the risk of online victimization.

By consulting statistical series produced in several countries, we intend to call into question such an assumption: do the figures really confirm this assertion? Is the evolution of the cybercrime trend correlated with the various phases of the health crisis?

In an attempt to answer those questions, CERT (Computer Emergency Response Teams) reports and police data will be used as the main data sources on the state of cybercrime, which will then be compared to data reflecting the changes in citizen lifestyles. The lockdown and restrictions on the mobility of individuals being one of the main indicators of these modifications, data on mobility is studied in depth in order to reconstruct the chronology of the lockdown periods.

The central question regarding the evolution of cybercrime trends during the health crisis will also be addressed by focusing on the evolution of cyberattacks on the international scene.

---

7. Daniel VENTRE, CNRS, CESDIP Laboratory (Guyancourt, France).

I.3.2. Chapter 2 – The SARS-CoV-2 pandemic crisis and the evolution of cybercrime in the United States and Canada[8]

As in the rest of the world, the pandemic crisis caused by SARS-CoV-2 in 2020 disrupted the normal functioning of societies in Canada and the United States. In terms of cybersecurity, it is highly probable for malicious actors to have adapted their practices to the pandemic context. This means that cybercrime evolved and became adapted to the new context. The chapter outlines these changes based on government and private organizations' reports on cybercrime, offering a critical perspective. Cybercrime in the United States and Canada during the pandemic crisis is analyzed in its various trends. The need for international cooperation to counter cybercrime and the methodological challenges encountered during the study conclude this chapter.

I.3.3. Chapter 3 – Online radicalization as cybercrime: American militancy during Covid-19[9]

The January 6, 2021 attack on the Capitol in Washington signaled the culmination of a broader period of sociopolitical activism in the United States. The trajectory of this period, which in many ways is still extending, closely followed the spread of the SARS-CoV-2 pandemic. The purpose of this chapter is to explore to what extent the basic analytical framework of cybercrime theory is still valid under pandemic conditions. We argue that the unprecedented pressure of accelerationism experienced in the United States during Covid-19 compels us to rethink online radicalization as a form of cybercrime. The extraordinary scope, speed and overall dynamics of accelerationist activity having challenged secular American institutions in recent years are all signs of a new kind of symbiotic association between online and offline elements.

---

8. Prof. Hugo LOISEAU, École de Politique Appliquée, University of Sherbrooke, Quebec, Canada.
9. Joseph FITSANAKIS, Professor of Intelligence and Security Studies, Coastal Carolina University, United States.
Alexa MCMICHAEL, Special Security Officer, Intelligence Operations Command Center, Coastal Carolina University, United States.

I.3.4. Chapter 4 – Cybercrime in Brazil after the Covid-19 global crisis: an assessment of the policies concerning international cooperation for investigations and prosecutions[10]

Cybercrime is not a new topic for the Brazilian authorities. Back in 2015, Brazil was the second most vulnerable country to cybercrime; Brazilian society having lost between US$4.1 billion and US$4.7 billion in data theft and financial fraud. How did cybercrime evolve during the Covid-19 pandemic period, and what responses did the government and Brazilian society provide to these issues? Based on quantitative data and interviews with members of the security forces, the chapter outlines the main lines of the reforms underway in the fight against cybercrime in Brazil.

I.3.5. Chapter 5 – Has Covid-19 changed fear and victimization of online identity theft in Portugal?[11]

This chapter focuses on online identity theft (OIT), now considered one of the fastest growing online crimes and leading to significant financial losses for victims. This research, undertaken in the Portuguese context (2017 and 2021), intends to (i) analyze the levels of victimization, fear and risk perception of online identity theft, before and after the Covid-19 pandemic crisis; (ii) study the evolution of routine online activities, both before and after the Covid-19 pandemic crisis; and (iii) understand the evolution of other forms of online victimization over the past 2 years.

I.3.6. Chapter 6 – A South African perspective on cybercrime during the pandemic[12]

This chapter examines cybercriminal activity during the Covid-19 pandemic from a South African perspective. Has South Africa been the

---

10. Alexandre VERONESE, Associate Professor of Social and Legal Theory at the Faculty of Law of the University of Brasilia, Brazil.
Bruno CALABRICH, Brazilian Federal Prosecutor. PhD candidate at the Faculty of Law of the University of Brasilia, Brazil.
11. Inês GUEDES, Joana MARTINS, Samuel MOREIRA, Carla CARDOSON, School of Criminology, Faculty of Law, University of Porto, Portugal.
12. Brett VAN NIEKERK, Durban University of Technology, South Africa.
Trishana RAMLUCKAN, University of KwaZulu-Natal, Durban, South Africa.
Anna COLLARD, KnowBe4 Africa, Cape Town, South Africa.

victim of an increase in cybercrime, or rather, is it the tactics of cybercriminals which have changed? There are divergent opinions on these issues. This chapter relies on exploratory research to study cybercrime trends in South Africa, while attempting to provide some answers to both questions. International rankings, national laws and regulations and incident reports are taken into account to analyze criminal trends. The lack of official reporting on cybercrime and cybersecurity incidents in South Africa is a limitation. The required information has been drawn from incident reports, industry reports and white papers illustrating trends.

I.4. References

[ALA 21] ALAVA S., "Internet est-il un espace de radicalisation ?" in MORIN D., AOUN S., AL BABA DOUAIHY S. (eds), *Le nouvel âge des extrêmes? : Les démocraties occidentales, la radicalisation et l'extrémisme violent*, Les Presses de l'Université de Montréal, 2021.

[ALT 07] ALTBEKER A., *A Country at War with Itself: South Africa's Crisis of Crime*, Jonathan Ball, Johannesburg and Cape Town, 2007.

[ANS 21] ANSSI, Organising a cyber crisis management exercise, available at: https://www.ssi.gouv.fr/en/guide/organising-a-cyber-crisis-management-exercise/, 2021.

[ANS 22] ANSSI, Panorama de la menace informatique 2021, Report 1.9.1, available at: https://www.cert.ssi.gouv.fr/uploads/20220309_NP_WHITE_ANSSI_panorama-menace-ANSSI.pdf, 2022.

[BAD 20] BADAWI E., GUY-VINCENT J., Cryptocurrencies emerging threats and defensive mechanisms: a systematic literature review, Faculty of Engineering, University of Ottawa, available at: https://ieeexplore-ieee-org.ezproxy.usherbrooke.ca/stamp/stamp.jsp?tp=&arnumber=9243940, 2020.

[BEC 92] BECK U., *Risk Society: Towards a New Modernity*, Sage, London, 1992.

[BEN 22] BENCHERIF A., BELPORO L.C., MORIN D., Étude internationale sur les dispositifs de prévention de la radicalisation et de l'extrémisme violents dans l'espace francophone, Chaire UNESCO en prévention de la radicalisation et de l'extrémisme violents, 2022.

[BOU 21] BOU SLEIMAN M., GERDEMANN S., "Covid-19: a catalyst for cybercrime?", *International Cybersecurity Law Review*, vol. 2, pp. 37–45, 2021.

[BRE 10] BRENNER S.W., *Cybercrime: Criminal Threats from Cyberspace*, Praeger, Santa Barbara, CA, 2010.

[CAR 18] CARASTATHIS A., SPATHOPOULOS A., TSILIMOUNIDI M., "Crisis, what crisis? Immigrants, refugees, and invisible struggles", *Refuge, Canada's Journal on Refugees*, vol. 34, no. 1, 2018.

[CHE 21] CHEN P., KURLAND J.R., PIQUERO A. et al., "Measuring the impact of the Covid-19 lockdown on crime in a medium-sized city in China", *Journal of Experimental Criminology*, pp. 1–28, 2021.

[CHI 21] CHIGADA J., MADZINGA R., "Cyberattacks and threats during Covid-19: a systematic literature review", *South African Journal of Information Management*, vol. 23, no. 1, pp. 1277, 2021.

[COL 20] COLLIER B., HORGAN S., JONES R. et al., "The implications of the Covid-19 pandemic for cybercrime policing in Scotland: a rapid review of the evidence and future considerations", *The Scottish Institute for Policing Research*, nos 1–18, available at: https://www.researchgate.net/publication/341742472_Issue_No_1_The_implications_of_the_Covid-19_pandemic_for_cybercrime_policing_in_Scotland_A_rapid_review_of_the_evidence_and_future_considerations, 2020.

[COR 20] CORDEY S., *The Evolving Cyber Threat Landscape during the Coronavirus Crisis*, Cyberdefense Project (CDP), Center for Security Studies (CSS), ETH Zurich, 2020.

[DEF 11] DEFLEM M. (ed.), "Introduction: criminological perspectives of the crisis", *Economic Crisis and Crime*, Emerald, Bingley, 2011.

[DUP 20] DUPONT B., "La cybercriminalité au temps de la Covid-19", *Policy Options*, available at: https://policyoptions.irpp.org/magazines/july-2020/la-cybercriminalite-au-temps-de-la-covid-19/, 2020.

[EUR 20a] EUROPOL, Corona crimes: suspect behind €6 million face masks and hand sanitisers scam arrested thanks to international police cooperation, available at: https://www.europol.europa.eu/media-press/newsroom/news/corona-crimes-suspect-behind-%E2%82%AC6-million-face-masks-and-hand-sanitisers-scam-arrested-thanks-to-international-police-cooperation, 2020.

[EUR 20b] EUROPOL, How Covid-19-related crime infected Europe during 2020, European Union Agency for Law Enforcement Cooperation, available at: https://www.europol.europa.eu/sites/default/files/documents/how_covid-19-related_crime_infected_europe_during_2020.pdf, 2020.

[FEA 09] FEARN-BANKS K., *Crisis Communications: A Casebook Approach*, Routledge, New York, 2009.

[GAY 21] GAYRAUD J.-F., "Les grandes criminalités, entre réalité géopolitique et menace stratégique", *Revue Défense Nationale*, vol. 7, no. 842, pp. 28–33, 2021.

[GOV 21] GOVENDER I., WATSON B.W.W., AMRA J., "Global virus lockdown and cybercrime rate trends: a routine activity approach", *Journal of Physics: Conference Series*, vol. 1828, 2021.

[GRY 21] GRYSZCZYNSKA A., "The impact of the Covid-19 pandemic on cybercrime", *Bulletin of the Polish Academy of Sciences Technical Sciences*, vol. 69, no. 4, 2021.

[GUP 21] GUPTA R., PRASAD A., BABU S. et al., "Impact of coronavirus outbreaks on science and society: insights from temporal bibliometry of SARS and Covid-19", *Entropy*, vol. 23, pp. 626, 2021.

[HAW 20] HAWDON J., PARTI K., DEARDEN T., "Cybercrime in America amid Covid-19: the initial results from a natural experiment", *American Journal of Criminal Justice*, vol. 45, pp. 546–562, 2020.

[HOR 21] HORGAN S., COLLIER B., JONES R. et al., "Re-territorialising the policing of cybercrime in the post-Covid-19 era: towards a new vision of local democratic cyber policing", *Journal of Criminal Psychology*, vol. 11, no. 3, pp. 222–239, 2021.

[IVA 12] IVASCHENKO O. et al., "The role of economic crisis and social spending in explaining crime in Russia", *Eastern European Economics*, vol. 50, no. 4, pp. 21–41, 2012.

[JAS 19] JASPERS J.D., "Business cartels and organised crime: exclusive and inclusive systems of collusion", *Trends in Organized Crime*, vol. 22, pp. 414–432, 2019.

[KEM 21] KEMP S., BUIL-GIL D., MONEVA A. et al., "Empty streets, busy internet: a time-series analysis of cybercrime and fraud trends during Covid-19", *Journal of Contemporary Criminal Justice*, 2021.

[KEN 21] KENNEDY L., SOUTHERN N.P., "The pandemic is putting gangsters in power, as states struggle, organized crime is rising to new prominence", *Foreign Policy*, 2021.

[KLE 14] KLEEMANS E.R., "Theoretical perspectives on organized crime", in PAOLI L. (ed.), *Oxford Handbook on Organized Crime*, Oxford University Press, Oxford, 2014.

[KON 97] KONTULA O., Crime in times of crisis, from research report summaries, Report, National Research Institute of Legal Policy, 1997.

[KOP 22] KOPPEL S., CAPELLAN J.A., SHARP J., "Disentangling the impact of Covid-19: an interrupted time series analysis of crime in New York City", *American Journal of Criminal Justice*, 2022.

[KOS 72] KOSELLECK R., "Krise", in BRUNNER O., KONZE W., KOSELLECK R. (eds), *Geschichtliche Grundbegriffe: Historisches Lexicon zur politisch-sozialen Sprache in Deutschland*, Klett-Cotta, Stuttgart, 1972.

[KUR 15] KURTZ J., Crisis and crime: examining the effect of macroeconomic conditions on criminal activity during the great recession, New York University, available at: https://as.nyu.edu/content/dam/nyu-as/politics/documents/Kurtz.pdf, 2015.

[LAA 21] LAAN J., The impact of the Corona-pandemic on the business model of cybercrime, Master's Thesis, University of Twente, 2021.

[LAZ 21] LAZAROV S., "The impact of Covid-19 on cybercrime trends", *Proceedings of the Sixth International Scientific Conference on Telecommunications, Informatics, Energy and Management*, TIEM 2021, pp. 62–65, 2021.

[MEY 21] MEYER C.O., BRICKNELL M., PACHECO P.R., How the Covid-19 crisis has affected security and defense-related aspects of the EU, European Parliament, Belgium, available at: https://www.europarl.europa.eu/RegData/etudes/IDAN/2021/653623/EXPO_IDA(2021)653623_EN.pdf, 2021.

[MIR 21] MIRÓ-LLINARES F., "Crimen, cibercrimen y Covid-19: desplazamiento (acelerado) de oportunidades y adaptación situacional de ciberdelitos", *Revista de los Estudios de Derecho y Ciencia Política*, IDP, no. 32, Marzo, Universitat Oberta de Catalunya, 2021.

[NAI 20] NAIDOO R., "A multi-level influence model of Covid-19 themed cybercrime", *European Journal of Information Systems*, vol. 29, no. 3, pp. 306–321, 2020.

[OEC 20] OECD, Coronavirus (Covid-19): joint actions to win the war, OECD Secretary General, available at: https://www.oecd.org/about/secretary-general/Coronavirus-Covid-19-Joint-actions-to-win-the-war.pdf, 2020.

[OLO 22] OLOFINBIYI S.A., "Cyber insecurity in the wake of Covid-19: a reappraisal of impacts and global experience within the context of routine activity theory", *Journal ScienceRise: Juridical Science*, vol. 1, no. 19, pp. 37–45, 2022.

[ONU 21] ONU INFO, La pandémie de Covid-19 a coûté 255 millions d'emplois en 2020 (OIT), available at: https://news.un.org/fr/story/2021/01/1087652, 2021.

[ONY 20] ONYEMA E.M., EUCHERIA N.C., OBAFEMI F.A. et al., "Impact of coronavirus pandemic on education", *Journal of Education and Practice*, vol. 11, no. 13, 2020.

[ORO 22] O'ROURKE R., *Covid-19: Potential Implications for International Security Environment – Overview of Issues and Further Reading for Congress*, Congressional Research Service, Washington, DC, 2022.

[PAR 20] PARK A., MONTECCHI M., FENG C. et al., "Understanding 'fake news': A bibliographic perspective", *Defense Strategic Communications*, vol. 8, pp. 141–172, 2020.

[PAY 20] PAYNE B., "Criminals work from home during pandemics too: a public health approach to respond to fraud and crimes against those 50 and above", *American Journal of Criminal Justice*, vol. 45, pp. 563–577, 2020.

[PLA 21] PLACHKINOVA M., "Exploring the shift from physical to cybercrime at the onset of the Covid-19 pandemic", *Journal of Cyber Forensics and Advanced Threat Investigations*, vol. 2, no. 1, pp. 50–62, 2021.

[SAN 22] SANS INSTITUTE, Glossary of security terms, available at: https://www.sans.org/security-resources/glossary-of-terms/, 2022.

[SMI 22] SMITH T., "Assessing the effects of Covid-19 on online routine activities and cybercrime: a snapshot of the effect of sheltering in place", *Caribbean Journal of Multidisciplinary Studies*, vol. 1, no. 1, pp. 36–60, 2022.

[TAB 20] TABREZ A., Corona virus (Covid-19) pandemic and work from home: challenges of cybercrimes and cybersecurity, available at: http://dx.doi.org/10.2139/ssrn.3568830, 2020.

[TRI 20] TRIPATHI K., Cybercrime against older people during Covid19 pandemic, UCL JDI Special Series on Covid-19, issue 4, available at: https://www.ucl.ac.uk/jill-dando-institute/sites/jill-dando-institute/files/cybercrime_0.pdf, 2020.

[UNO 12] UNODC, Monitoring the impact of economic crisis on crime, Vienna, available at: https://www.unodc.org/documents/data-and-analysis/statistics/crime/GIVAS_Final_Report.pdf, 2012.

[VEN 21] VENKATESHA S., RAHUL R.K., CHANDAVARKAR B.R., "Social engineering attacks during the Covid-19 pandemic", *Springer Nature Computer Science*, vol. 2, no. 78, 2021.

[WAN 20] WANG M.W., ZHOU M.-Y., JI G.-H. et al., "Mask crisis during the Covid-19 outbreak", *European Review for Medical and Pharmacological Sciences*, vol. 24, pp. 3397–3399, 2020.

[YU 21] YU Y.J., PARK Y.S., KELLER A. et al., "A mixed methods systematic review of the impacts of coronavirus on society and culture", *International Journal of Environmental Research and Public Health*, vol. 18, pp. 491, 2021.

1

The Evolution of Cybercrime During the Covid-19 Crisis

1.1. Introduction

Since the early months of 2020, several publications have addressed the impacts of the pandemic on the evolution of cybercrime.

In March 2020, the US Congressional Research Service (CRS) published a brief report on the subject [FIN 20]; in April 2020, UNODC warned against the risks specifically related to the new pandemic context [UNO 20]; followed by the French Senate [SÉN 20] during the same month. In June 2020, the international organization Global Initiative against Transnational Organized Crime [MAH 20], the German Konrad-Adenauer Foundation [WIG 20] and Interpol [INT 20] (by means of a survey studying 192 countries) produced numerous reports, whose analyses converged on various points:

– the pandemic was a context favoring the acceleration of cybercrime;

– during lockdown, an increase was observed in crimes in cyberspace, both as "cyber-dependent crimes" and "cyber-related crimes";

– a considerable increase in cybercriminal activities related to Covid-19;

Chapter written by Daniel VENTRE.

For a color version of all figures in this chapter, see http://www.iste.co.uk/ventre/cybercrime.zip.

– in addition to the increase in the level of cybercrime, the pandemic may also have disrupted the security forces and the justice system in charge of fighting this phenomenon;

– the evolution of cybercrime varies from one world region to another (according to Interpol);

– the adaptation of cybercriminals to the new environment, who change their tactics in order to maximize profits: "Cybercriminals are shifting their focus from individuals and small businesses to large corporations, governments and critical infrastructure which play a fundamental role in the response to the disease outbreak" [INT 20].

These official reports were supplemented by the work of university researchers who observed and attempted to explain the processes and phenomena at work concerning the evolution of cybercrime in this particular pandemic context. Their analyses notably focused on cybercrime in its "national" sphere. Much research has been published about cybercrime in Indonesia [KAS 20], China [CHE 20], Poland [GRY 21], the United Kingdom [BUI 20], the Maldives [WAH 21] and the United States [HAW 20], various readings which envision cybercrime not only as a global, transnational phenomenon, but as an issue in which the local dimension is essential. These considerations fuel reflection on the evolution of national legal norms (as opposed to international norms) and the local/national dimension of the security organization (police, justice) [HOR 21].

Cybercrime is thought to have been particularly "effective" or active due to the convergence of several factors: the online presence of a greater number of Internet users, an increase in insecure practices during lockdown periods, reduced vigilance and less presence of cybersecurity actors, opportunistic criminal actions through targeted and thematized attacks. During the lockdown period, citizens spent longer hours online, exposing themselves to more risks due to their poorly or insufficiently secure practices (teleworking, e-commerce, online games, social media, etc.). The increased use of e-commerce provided cybercriminals with more opportunities to steal banking and personal data or carry out fraudulent actions. Figures from the UK Office for National Statistics[1] identified the growth of e-commerce since

---

1. Available at: https://www.ons.gov.uk/businessindustryandtrade/retailindustry/timeseries/j4mc/drsi.

the onset of the pandemic (bearing in mind that this is only an example and no generalizations can be made regarding the evolution of e-commerce in the world). Growth was particularly remarkable for the year 2020, the year 2021 presenting a slight decline in e-commerce activity, even if the volumes remained much higher than before the health crisis.

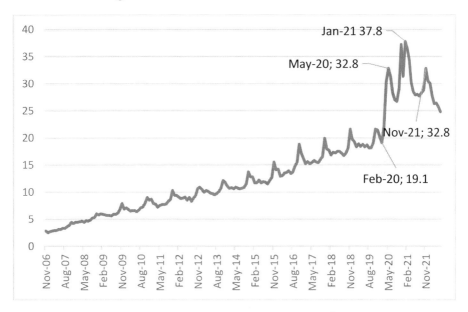

Figure 1.1. *Internet sales as a percentage of UK total retail sales, November 2006 to May 2022. Figure reconstructed from data published on the site: https://www. ons.gov.uk/businessindustryandtrade/retailindustry/timeseries/j4mc/drsi*

Statistics from the International Telecommunication Union (ITU)[2] show that the number of Internet users continued to grow in 2020 and 2021, though without indicating a remarkable acceleration.

While growth is continuous – regardless of the level of economic development of the states – the gaps between industrialized countries and the least developed countries substantially persist. The pandemic period left this scenario unchanged. At present, barely more than 60% of the population has access to the Internet at a global scale. This also means that the issues

---

2. Available at: https://www.itu.int/en/ITU-D/Statistics/Pages/stat/default.aspx.

discussed in this chapter (exposure to cyber risks, cybercrime, impact of crises on cybercrime and cyberspace) are not yet universal issues. A significant part of humanity remains unaffected by such considerations.

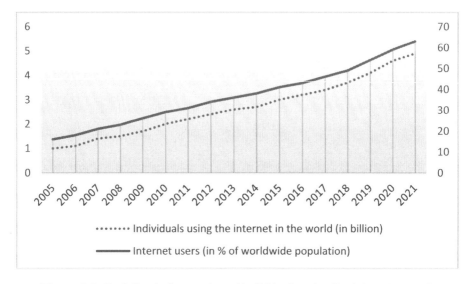

Figure 1.2. *Evolution in the number of individuals using the Internet around the world (reconstructed from data published by the ITU on the site: https://www.itu.int/en/ITU-D/Statistics/Pages/stat/default.aspx)*

1.2. Observing the evolution of cybercrime

The guiding question to be analyzed will be: Is there a correlation between the evolution in cybercrime trends and the various phases of the pandemic?

This chapter postulates that the evolution of cybercrime can be explained, at least in part, by changes in the uses of cyberspace, themselves resulting from a change in lifestyle imposed by the pandemic and the various national measures enforced to manage the health crisis. Two types of data will be compared: those relating to the evolution of cybercrime and contextual data (the restrictions imposed on citizen mobility being one of the essential markers of lifestyle changes occurring during this period, for which mobility data will be leveraged).

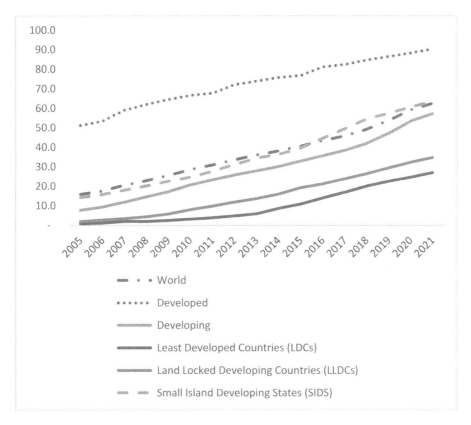

Figure 1.3. *Rate of Internet use in the population (reconstructed from data published by the ITU[3])*

The data on "cybercrime" used in this study have been drawn from CERTs[4] (number of cyber incident reports received and recorded) and police statistics (number of complaints received, cases processed, individuals arrested). These organizations publish statistics, whose frequency can be annual, monthly, sometimes weekly, and more rarely, daily. The level of detail of the information published is not internationally standardized, which means that the categories of cyber incidents and cybercrimes or offenses taken into account are not identical everywhere. The figures published by these organizations, by means of annual reports or via their websites, each

---

3. Available at: https://www.itu.int/en/ITU-D/Statistics/Documents/facts/ITU_regional_global_Key_ICT_indicator_aggregates_rev1_Jan_2022.xlsx.
4. CERT: Computer Response Emergency Team.

offer a particular perspective on cybercriminal activity affecting society in a particular country. The level of development of these official statistics is uneven depending on the country, and on the world region, certain series of data offer a panoramic view of the last two decades, while others have been compiled only recently.

For this first part of the study, data have been sorted into three categories, depending on the frequency of data available: annual, monthly and weekly.

Table 1.1 summarizes the data sources used.

Country	Global Cybersecurity Index (ITU) (2020)[5]	Period covered	Annual or monthly data	Source (CERT, Police, etc.)
India	10	2002–2020	Annual	CERT and Ministry of Home Affairs
Brazil	18	1999–2020	Monthly	CERT.br
Malaysia	5	January 1998–June 2022	Monthly	CERT[6]
Northern Ireland (United Kingdom)	2	2017–2022	Monthly	Police
China	33	2017–2022	Weekly	CNCERT

Table 1.1. *Summary of the cybercrime data sources used in this chapter*

Some variables have been chosen to identify the periods during which lifestyle changes may have been particularly significant due to mobility restrictions, as well as a set of measures dictated by governments or the local authorities for managing the health crisis (social distancing, reduced opening hours at the shops, restriction of commercial activity, etc.). The following data sources will be used:

---

5. Available at: https://www.itu.int/dms_pub/itu-d/opb/str/D-STR-GCI.01-2021-PDF-E.pdf.
6. Available at: https://www.mycert.org.my.

Variable	Available data time frame	Source
PHSM (Public Health and Social Measures) Severity Index	Daily	Two indicators: "Stringency index", "Health and containment index", produced by researchers at the University of Oxford
Mobility	Daily	Data from Google Mobility[7]. The data available at the time of this study span from February 2020 to July 2022

Table 1.2. *List of variables confronted with data on cybercrime[8]*

Google has uploaded files describing the mobility of the global community since the onset of the pandemic in February 2020 in csv format[9]. These data reveal a daily measurement, per country, of the attendance to certain categories of places: shops and leisure areas, grocery stores and pharmacies, parks, public transport stops, workplaces and places of residence.

These data, as well as the reports published on the Google site[10], "are intended to provide information on changes brought about by policies to combat Covid-19". Above all, they provide a fairly accurate picture of the traveling restriction periods. These are reflected, for example, by the sharp decrease in the attendance to workplaces or, on the contrary, the increased presence of individuals in residential spaces.

Mobility is one of the remarkable aspects of how lifestyles change during a pandemic. But are mobility restrictions correlated with an increase in cybercrime?

The "stringency index", created by a team from the University of Oxford as part of the "Oxford Coronavirus Government Response Tracker (OxCGRT)" project is a composite measure built on the basis of nine pandemic response indicators. These indicators include:

– school closures;

– workplace closures;

– cancellation of public events;

---

7. Available at: https://www.google.com/covid19/mobility/?hl=fr.
8. For this report, the data are used in the Malaysia case study.
9. Files can be downloaded from: https://www.google.com/covid19/mobility/.
10. Available at: https://www.google.com/covid19/mobility/.

– restrictions on public gatherings;

– closures of public transport;

– stay-at-home requirements;

– public information campaigns;

– restrictions on internal movements;

– international travel controls.

The index has a value between 0 and 100 (100 = strictest level). The index is included in the database of daily Covid cases (new cases, deaths), for all countries around the world, downloadable from ourworldindata.org[11]. The data cover the period from February 2020 to July 2022. This index has been enriched with four additional indicators relating to:

– the testing policy;

– the extent of contact tracing;

– face coverings;

– vaccine policy.

This index, a composite measure of 13 indicators, is called the "Health and Containment Index". An index value is assigned to each country daily that is also between 0 and 100 (100 = strictest level). The data analyzed focus on the period between February 2020 and July 2022. The index builds on the above-mentioned database (.cvs file).

Some of the indicators used to calculate the index are publicly available, which makes it possible to isolate specific aspects from the various measures taken during the health crisis management[12].

1.2.1. *Leveraging annual data: the case of India*

While annual data may seem insufficient to discuss the hypotheses on the close link between the teleworking period and the increase in cybercrime, for instance, or to cross-reference the observed changes in cybercrime

---

11. Available at: https://ourworldindata.org/covid-stringency-index#learn-more-about-the-data-source-the-oxford-coronavirus-government-response-tracker.

12. Available at: https://ourworldindata.org/covid-stay-home-restrictions.

phenomena with any other event or potentially explanatory variable, such data may still prove invaluable.

In particular, they make it possible to observe changes in the long run, not only within a reduced time frame (2020–2022): what about the impact of the pandemic period on the general evolution of cybercrime trends?

The data published by the Indian Ministry of Home Affairs include:

– the number of registered cases;

– the number of individuals arrested.

Data are provided for each state and distributed via the two main legal corpuses: the Information Technology (IT) Act and the Indian Penal Code (IPC). For each corpus, the number of cases is distributed following each cybercrime category. Demographic data are also provided (age brackets of the individuals arrested).

Those data offer a view on the geography of cybercrime and its demographic aspects (age brackets and the evolution of age categories in time and space). The analyst can also deduce the average number of individuals involved per case handled, capture the most widespread forms of crime and detect which legal corpus is the most often summoned in the fight against cybercrime (IT Act versus IPC).

At the time of writing these lines, the series of reports stops at the year 2020, which only offers a partial vision of the developments that may have occurred during the pandemic period. However, the level of detail of the available data may be deemed sufficient for a preliminary analysis.

If the official statistics are to be credited, the first year of the pandemic was not accompanied by a radical increase in cybercrime. For all categories, growth started taking shape around 2010. The number of cases falling under the IT Act, the main corpus concerning cybercrimes, even recorded a slight decline in 2020. In this case, it may be interesting to access monthly data, which could provide a more accurate assessment of the link between the various phases of the pandemic, its impact on society and the (concomitant) evolution of cybercrime.

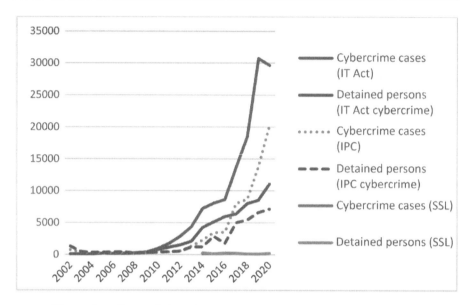

Figure 1.4. *The evolution of cybercrime in India. Curves reconstructed from data available on the site: https://ncrb.gov.in/en/crime-india*

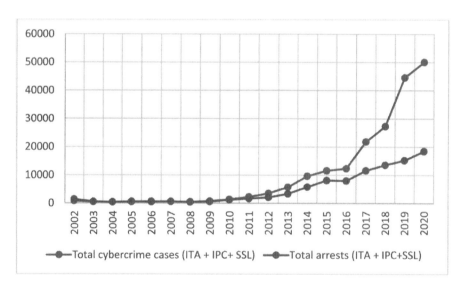

Figure 1.5. *Evolution of cybercrime in India (number of cases, number of arrests). Curves reconstructed from data available on the site: https://ncrb.gov.in/en/crime-india*

It is interesting to emphasize that the activity of the police forces in 2020 did not suffer from the effects of the epidemic (which could have been disruptive for the security sector, to the same extent it had affected all the other sectors: police, military, company employees, without distinction, all were exposed to disease lethality and morbidity). The publication of data for 2021 would be a crucial indicator for assessing the capacity of security actors to maintain their activity in degraded conditions.

The statistics of the Indian national CERT also reflected an increase in cybercrime in 2020 (about three times more incidents reported than in 2019):

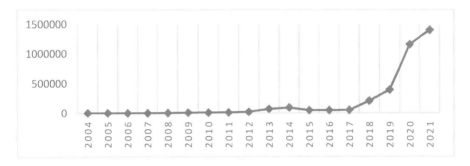

Figure 1.6. *Number of cyber incidents reported to CERT.in (reconstructed from CERT reports)*

While the increase in phenomena was remarkably strong in 2020 (and persisting in 2021), it should be noted that 2018 marked the key moment in the trend reversal, with the phenomena having remained relatively stable until then.

1.2.2. Leveraging monthly data

1.2.2.1. Malaysia

Throughout the period 1998–2022, the typology of cyber incidents recorded by CERT Malaysia, based on the reports received, comprises 13 categories. It is important to note that the typology varies over time, and that the data series do not always cover the entire 1998–2022 period.

Cyber incidents recorded by Malaysia CERT	Period
Cyber harassment	From January 1998 to December 2021
Content related/indecent content	Introduced in January 2009
Fraud/forgery	Over the entire 1998–2022 period
Spam	Over the entire period, but no data from January 2005 to December 2009
Intrusion attempt/vulnerability probing	Introduced in January 2009
Malicious codes	Introduced in January 2005
Viruses	From January 1998 to December 2004
Mail bomb	From January 1998 to December 2004
Denial of service	Over the entire 1998–2022 period
Intrusion	Over the entire 1998–2022 period
Vulnerabilities report/drone report	Introduced in January 2009
Hack threat	From January 1998 to December 2008
Destruction	From January 1998 to December 2004

Table 1.3. *Typology of cyber-incidents recorded by Malaysia CERT (1998-2022)*

Finding 1.1: For the January 2020–June 2022 period, the total number of incidents reported monthly peaked in April 2020 and was immediately followed by a period of decline. However, over the entire period considered, the (exponential) trend curve translated a significant decrease in phenomena as a whole.

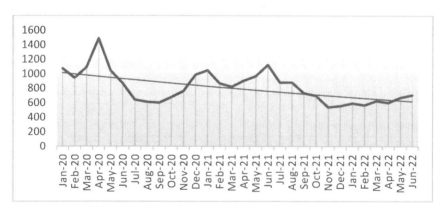

Figure 1.7. *Total of cyber incidents reported to the CERT in Malaysia, from January 2020 to June 2022 (curve reconstructed from data available on the site: https://www.mycert.org.my)*

Finding 1.2: The downward trend observed over the 2020–2022 period alone is set against a broader long-term upward trend.

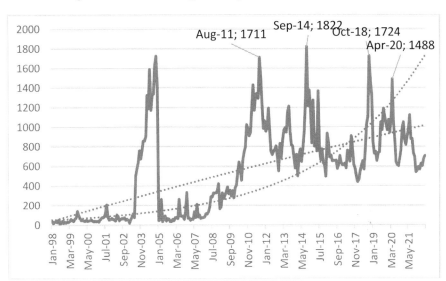

Figure 1.8. *Evolution in the number of cyber incidents reported to the CERT in Malaysia, from January 1998 to June 2022 (curve reconstructed from data published on the site: https://www.mycert.org.my)*

The spike of the years 2003 and 2004 essentially included "spam" reports. This category disappeared from CERT statistics in January 2005.

The April 2020 spike was not particularly outstanding: there have been others, even higher ones, over the last decade.

Finding 1.3: Fraud occupied a prominent place in all the reported incidents. It also largely contributed to the April 2020 spike. Intrusions and malware came in second. The other categories of cybercrime were practically marginal.

Finding 1.4: The curve of the number of cyber incidents practically showed the same shape and the same pace as the evolution of the mobility index in residential zones.

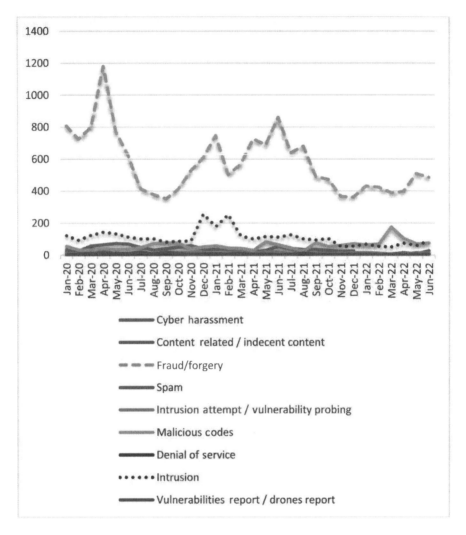

Figure 1.9. *Cyber incidents reported to the CERT in Malaysia, from January 2020 to June 2022 (curve reconstructed from data published on the site: https://www.mycert.org.my)*

Finding 1.5: The shape of the cyber incident curve also followed that of the two indexes very closely ("stringency index" and "health and containment index").

The Evolution of Cybercrime During the Covid-19 Crisis   15

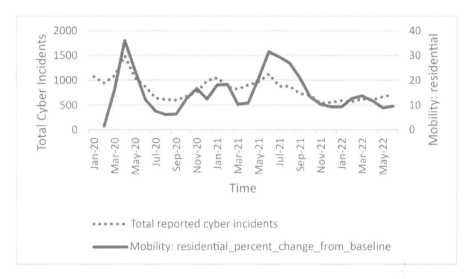

Figure 1.10. *The two curves (cyber incidents reported, evolution of the mobility index in residential zones) have practically the same shape (curve reconstructed from data published on the site: https://www.mycert.org.my)*

Figure 1.11. *Curve of cyber incidents versus "stringency" and "health and containment" indexes (curve reconstructed from data published on the site: https://www.mycert.org.my)*

Finding 1.6: The evolution curve of the mobility index at the workplace clearly emphasizes the lockdown and teleworking periods at the national level. When the index decreases, it means that fewer people are moving to the workplaces. The cyber incident reporting curve evolved at the same pace, but in a diametrically opposite direction: the lower the mobility to the workplace, the higher the number of reported cyber incidents. This relationship remained constant throughout the 2020–2022 period.

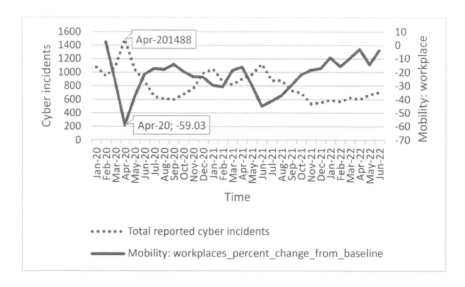

Figure 1.12. *The number of reported cyber incidents is negatively correlated with the evolution of the mobility index (workplace) (curve reconstructed from data published on the site: https://www.mycert.org.my)*

This relationship is even more marked when comparing the evolution of cyber incidents and mobility in essential businesses (groceries, pharmacies).

The Malaysian Communications and Multimedia Commission (MCMC) also records complaints from Internet users related to content. The published figures show a significant increase in the number of reports, between 2019 and 2020, of around 100%. The complaints are classified as follows:

Figure 1.13. *Evolution in the number of cyber incidents versus mobility index (essential services) (curve reconstructed from data published on the site: https://www.mycert.org.my)*

Elements of complaints	2019	2020	Variation (%, 2020 as compared to 2019)
Obscene	850	1,637	+92.6
False	3,050	6,637	+117.6
Offensive	2,312	4,535	+96.2
Indecent	188	373	+98.4
Menacing	88	151	+71.6
Other	3,938	7,472	+89.7
Total	**10,426**	**20,805**	**+99.5**

Table 1.4. *Evolution in the number of complaints relating to content on the Internet. Table reproduced from data on the MCMC (Malaysian Communications and Multimedia Commission) and published on the dosm.gov.my[13] website*

---

13. Available at: https://www.dosm.gov.my/v1/index.php?r=column/cthemeByCat&cat=455&bul_id=eHE0eGZWSmNROG1BbHR2TzFvZzZxQT09&menu_id=U3VPMldoYUxzVzFaYmNkWXZteGduZz09.

Such an increase is only visible in certain categories of computer crimes. While annual statistics have the disadvantage of erasing the details, on the more positive side, they make it possible to attach less importance to them. The evolution curve of CERT Malaysia statistics in terms of annual data presents a different perspective on the 2020–2021 period. The year 2020, at the same level as 2019 and 2021, even shows a slight decline. The pandemic years are not coincident with an upsurge in cybercrime acts (at least those taken into account by CERT).

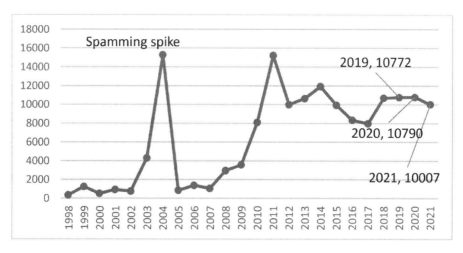

Figure 1.14. *Annual evolution in the number of cyber incidents recorded by the CERT of Malaysia (reconstructed from CERT statistics)*

Interpreting CERT data alone, the pandemic period did not produce the drastic increase in cybercrime that many reports and articles around the world reported. It therefore becomes imperative to draw a distinction between specific cybercriminal phenomena having shown a sharp increase (online fraud in Penang is reported to have multiplied by four in the space of one year, between 2019 and 2020, according to police data. The phenomenon could be explained by the increase in online shopping volumes, exposing consumers to the risk of fraud) [BAS 20], longer term trends and the evolution of overall cybercriminal acts.

1.2.2.2. *Brazil*

The shape of the incident curve reported to CERT only slightly differs from the available data for Malaysia. First, it is worth recalling that before entering the pandemic period, cybercrime had its own specificities in each country, its unique evolution, a chronology and a specific history, and this despite the fact that cyberspace is an inter or transnational space, and that cybercrime can be deployed therein without the constraints of territorial borders. The national, endogenous dimension is central because cybercrime or delinquency does not always manifest at an international scale. On the other hand, the pandemic evolved at significantly different paces from one continent to another, and even from one country to another in the same region. The policies accompanying this epidemic differed intently from country to country, the populations around the world did not all experience lockdown at the same time, teleworking was not imposed everywhere in the same way, schools were not closed in all places for the same time lapse, etc. Societies did not face the pandemic in the same way, and their lifestyles, or their economies, were not affected in the same way.

Unlike the Malaysian case (and various other countries), the evolution in the number of cyber incidents reported to CERT.br, on the basis of monthly data, does not show a spike in April 2020, but a first spike in March, the date after which the volume of incidents remains more or less stable, and high, around 60,000 cases/month, against less than 40,000 in January for the same year.

Figure 1.15. *Evolution in the total number of cyber incidents recorded by CERT.br in 2020*

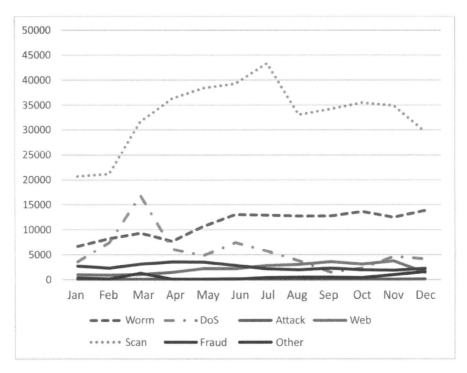

Figure 1.16. *Number of cyber incidents reported to CERT.Br in 2020 (graph reconstructed from the CERT.br annual report)*

Cyber scanning is the type of major incident reported in Brazil, followed by virus attacks (worms) and distributed denial-of-service (DDoS) attacks. The evolution of each type of cyber incident is characterized by its own curve: scanning practices increased in intensity in March 2020, then continued to grow until July; viral attacks were part of a more linear progression, with no real moment of greatest intensity. As for denial of service attacks, they were more virulent in March 2020, but, on the whole, tended to decline throughout the following months.

The CERT.br report provides a separate table for the data concerning spam, which on its own, exceeds all other categories of cyber incidents. The spike of activity was recorded in May 2020, but its progression had been constant since January.

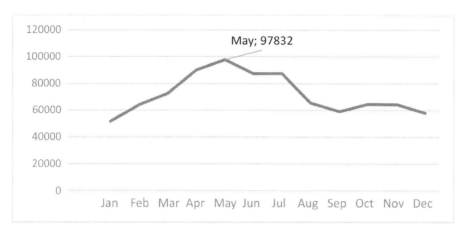

Figure 1.17. *Evolution in the number of spamming cases recorded by CERT.br during the year 2020 (curve reconstructed from CERT.br data)*

As in the case of Malaysia, one type of cyber incident dominates and shapes the curve of the total reported incidents. But by examining the evolution of each category of incident in detail, it becomes evident that phenomena follow trajectories which are sometimes quite different from one another.

1.2.3. *Leveraging weekly data: the case of China*

The data used for monitoring the evolution of cybercrime in China have been drawn from CNCERT. They have been published weekly since the end of July 2017. These weekly data involve the following variables:

– the number of infected computers in China (controlled by Trojans or botnets);

– the number of defaced websites in China;

– the number of backdoored websites in China;

– the number of phishing web pages targeting sites in China;

– the number of new vulnerabilities (including low-, medium- and high-risk vulnerabilities);

– the number of network security incidents handled by the CNCERT.

As Google does not publish mobility data for China, the "stay-at-home requirements"[14] index will be used instead. It provides a daily assessment of the level of mobility restrictions imposed by the Chinese authorities [HAL 21]. The authors of this index have split it into four levels:

0: absence of restrictive measure;

1: recommendation not to leave the house;

2: leaving the house is forbidden, except in special cases (daily exercise, food shopping, essential travel);

3: leaving the house is strictly forbidden, with only a few exceptions (only going out once in a given period, only one person going out at a time, etc.).

The exceptions have been nuanced in various ways depending on the country, sometimes the region, or on a case-by-case basis. Here, only the note providing an overall picture of the restriction threshold will be retained. In China, the highest level of restriction occupied a significant portion of the timeline starting in January 2020. This does not mean that all of China was under strict lockdown during the whole period, because lockdown policies were often managed at a local level.

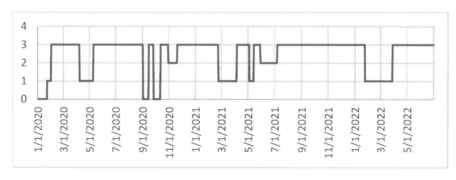

Figure 1.18. *Evolution of the "stay-at-home requirements" index for China, from January 1, 2020 to May 29, 2022. Reconstructed from data published on the site: https://ourworldindata.org/grapher/stay-at-home-covid*

---

14. Available at: https://ourworldindata.org/grapher/stay-at-home-covid.

Since the index data were calculated on a daily basis, it was also necessary to estimate a weekly average to be comparable with the CNCERT information.

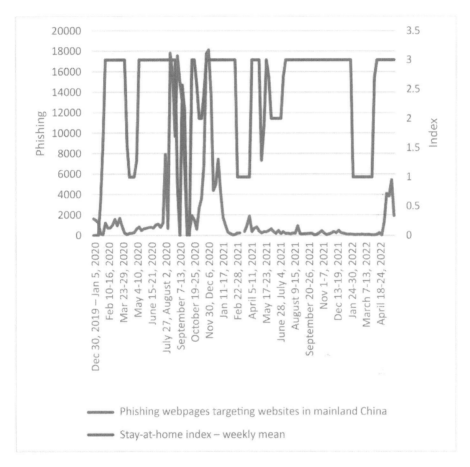

Figure 1.19. *Respective evolution of the "stay-at-home" index and the number of phishing sites detected in China since January 2020*

During lockdown periods, phishing attacks were particularly numerous. There were spikes in cybercriminal activity, with nearly 18,000 phishing sites identified, and a hyperactivity observed from August until the end of December 2020. The second half of the year was particularly characterized by phishing, although such operations had already been detected between

January and July 2020 (e.g. actions of APT 32, of Vietnamese origin, within the framework of espionage operations targeting healthcare actors and Chinese authorities)[15].

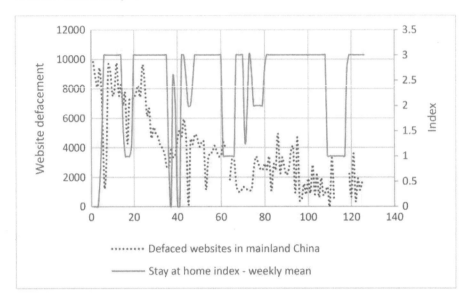

Figure 1.20. *Respective evolution of the "stay-at-home" index and the number of defaced websites detected in China since January 2020*

Cyberattacks defacing websites are less and less frequent. This type of criminal activity, often politicized, tends to decrease everywhere in the world. China's situation is part of this global dynamic.

Like others, the "backdoor infected websites" variable evolved independently, not being correlated with periods featuring lifestyle changes (lockdown and social restriction), and following a clearly downward trend. This decrease could be interpreted as the strengthening of Chinese cybersecurity capabilities, successfully preventing attacks against its servers. The pandemic seems not to have affected its cybersecurity capabilities. But this observation contradicts the previous one, according to which the number of infected computers keeps increasing constantly.

---

15. Available at: https://www.mandiant.com/resources/apt32-targeting-chinese-government-in-covid-19-related-espionage.

The Evolution of Cybercrime During the Covid-19 Crisis 25

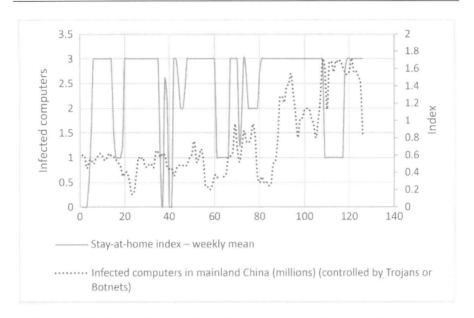

Figure 1.21. *Respective evolution of the "stay-at-home" index and the number (in millions) of infected computers (Trojan horses and botnets) in China since January 2020*

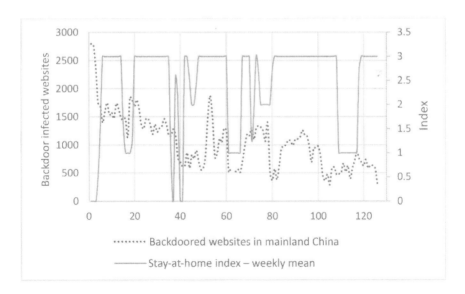

Figure 1.22. *Respective evolution of the "stay-at-home" index and the number of new backdoor infected websites in China since January 2020*

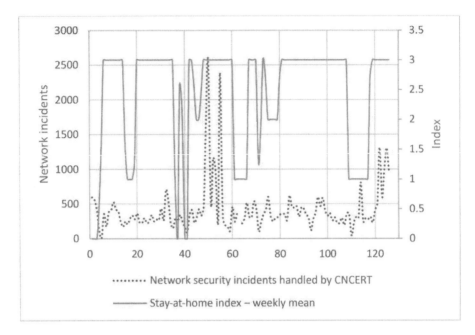

Figure 1.23. *Respective evolution of the "stay-at-home" index and the number of network security incidents managed by the CNCERT since January 2020*

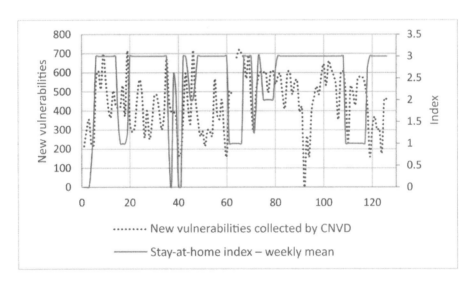

Figure 1.24. *Respective evolution of the "stay-at-home" index and the number of new vulnerabilities identified by the Chinese National Vulnerability Database (CNVD) since January 2020*

This variable (number of network security incidents) evolved independently from the lifestyle changes induced by lockdown policies. During these periods, citizens were required to stay at home, and to rely on networks for communication and work. We could have expected the number of security incidents to increase proportionally, but apparently this was not the case.

The volume of new vulnerabilities detected is relatively stable, in a band between 300 and 600 new vulnerabilities/week, with no major spikes. This indicator evolved autonomously. We should simply retain that the lockdown or social restriction periods did not affect the action of cybersecurity actors responsible for network monitoring.

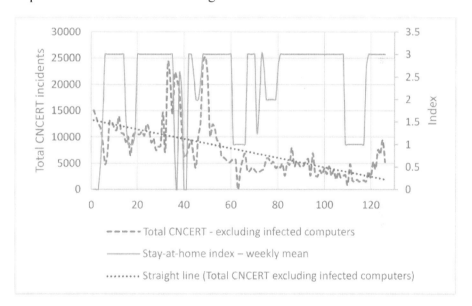

Figure 1.25. *Respective evolution of the "stay-at-home" index and overall incidents/vulnerabilities (excluding computers infected by botnets and Trojans) recorded by CNCERT since January 2020*

Over the January 2020–May 2022 period, the total number of incidents recorded by the CNCERT decreased. However, this period was marked by an upsurge in cybercriminal activity and frequent variations were observed in the mobility index between August 2020 and January 2021.

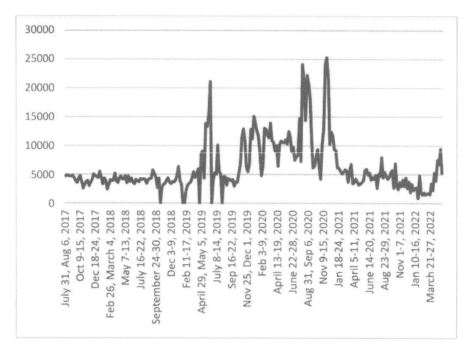

Figure 1.26. *Respective evolution of the "stay at home" index and overall incidents/vulnerabilities (excluding computers infected by botnets and Trojans) recorded by the CNCERT since July 2017*

The evolution of the total incidents recorded by the CNCERT since 2017 (excluding botnet infected computers), apart from the spike recorded in the middle of 2019, was part of a relatively stable trend (slightly below 5,000 incidents weekly). The pandemic period featured a significant increase in the number of weekly incidents (ranging between 5,000 and 15,000), throughout the year 2020 and at the beginning of the year 2021. But this acceleration was rooted in October–November 2019, and probably even much earlier, around May 2019 if we include the mid-year spike in activity and not the epidemic at the end of December 2019/beginning January 2020. Perhaps, this could suggest a dynamic which cannot be explained solely by lifestyle changes and the Internet uses of Chinese citizens and businesses. The variations observed during the pandemic period cannot be explained by the health crisis context alone. It would therefore be advisable to focus on other possible driving sources underlying this dynamic: in particular, international tensions in the Asia-Pacific zone or political and economic tensions with the American competitor, for example.

1.3. Has the global geography of cyberattacks changed?

This section will examine the geographical distribution of cyberattacks in the world, comparing the pre-pandemic scenario with the pandemic itself. Cyberattacks are offensive, aggressive actions or operations perpetrated by one or more actors in cyberspace, targeting and affecting one or more individuals/organizations or systems. Not only do cyberattacks include organized crime actions, but also operations carried out by the states themselves, such as espionage or even sabotage. Sometimes the two universes cooperate. Were there any significant changes in the geographical distribution of cyberattacks over the 2020–2022 period? To provide some answers, the data published by the CSIS (Center for Strategic & International Studies) will be used. Since 2006, this institution has been supplying a list of major cyber incidents taking place around the world. Of course, the list does not claim to be exhaustive.

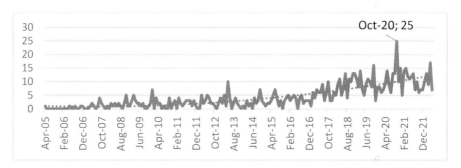

Figure 1.27. *Number of cyber operations recorded monthly by the CSIS from April 2005 to May 2022*

The number of cyberattacks recorded by CSIS grew steadily, albeit at a relatively modest rate in the early years.

There was a clear acceleration in 2017 (23 more operations, that is, +56% compared to 2016) and 2018 (44 more operations, that is, +68% compared to 2017). The increase in the number of attacks recorded continued to progress until 2020 (+ 22% compared to 2019), but 2021 was marked by a slight decrease. It is still unclear whether this trend will be confirmed by the end of

2022. In 2020, the increase was less acute than in the previous two years, even if it embodied the highest point in the histogram. The year 2020 seemed to act more as an extension of a dynamic initiated two years earlier, but not as an amplifier. The pandemic did not constitute a particular context for the amplification of the cyberattack phenomenon at a planetary scale.

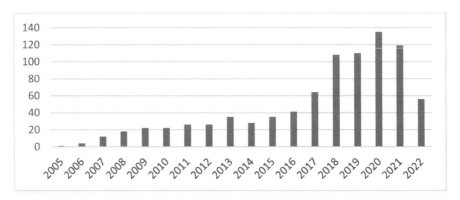

Figure 1.28. *Number of cyber operations recorded yearly by the CSIS from April 2005 to May 2022*

The first remarkable point, though, is the order of magnitude: it is not a question of thousands, not even less, of hundreds of thousands of attacks or cyber incidents. During the 2020 peak of activity, cyberattacks barely exceeded 120 units. This can be due to at least two reasons: the data collection task by the CSIS is not and cannot be exhaustive (how can we claim to identify all the cyberattacks that are taking place across the globe?); cyberattacks are operations which result in tens, hundreds and sometimes even millions of hit "targets" (i.e. computers, servers, data, etc.).

The driving forces or contexts of these attacks are multiple: political (security and defense issues, armed conflicts, international crises, rivalries, power politics) and economic (international interstate competition, economic power, etc.). Statistics based on data published by the Council on Foreign Relations highlight the various driving sources of cyberattacks.

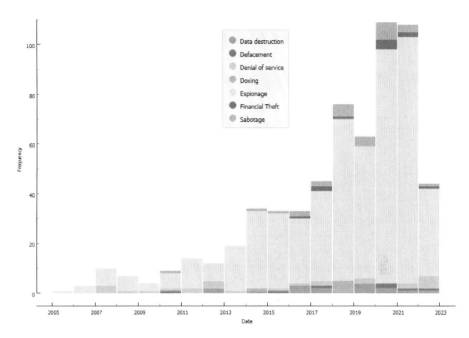

Figure 1.29. *Cyberattacks – CFR data – annual frequency for each attack motive (histogram reconstructed from data published on the CFR website, June 9, 2022)*[16]

An examination of the geographical distribution of cyberattacks, according to CSIS data, sheds light on the confrontation formats and the hackers' perimeters of action, both at the state level and regarding organized cybercrime.

From 2020 to 2022, 19 states were identified at the origin of the attack, or expressly designated as aggressors, and 87 states/world regions or international organizations (e.g. NATO, UN) as targets or victims of attacks. For approximately 30% of the operations referenced, no mention of the geographical origin of the attack has been indicated.

A cluster of a few countries has been systematically identified at the origin of the attacks:

– for the period before 2020, China is on top of the list, followed by Russia, Iran and North Korea. A set of consequential attacks does not

---

16. Available at: https://www.cfr.org/interactive/cyber-operations/export-incidents?_format=csv.

mention any particular attribution (30% of the list of cyberattacks in 2005–2019);

– for the 2020–2022 period, China is again on top of the list, still followed by Russia, Iran and North Korea. Next comes the United States, but in a much smaller proportion (10 times less than China in particular). The list follows with 14 other states or regions.

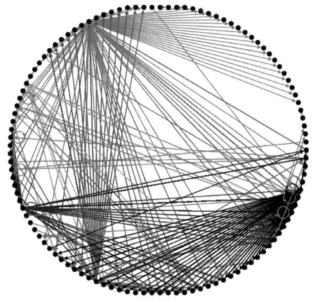

Figure 1.30. *Distribution of cyberattacks worldwide, according to CSIS reports, for the period 2005–2019. The arrows point from the attack origin toward the targets/victims (in black, attacks from China; in red from Russia, in dark blue from Iran; in orange, from North Korea)*

As for the victim or target countries:

– For the period before 2020, the United States ranked first, followed by Germany and South Korea.

– For the period 2020–2022, the United States was still at the top of the most affected states. Then came Ukraine, Israel, the United Kingdom, India, Russia, Iran, and after those, 70 other states, regions or international organizations.

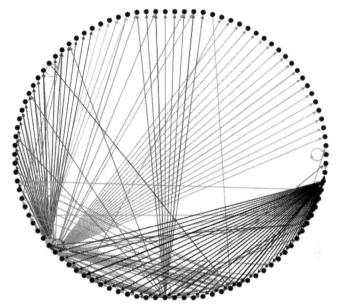

Figure 1.31. *Distribution of cyberattacks around the world, according to CSIS reports. Period analyzed on the graph: 2020–2022. The arrows are oriented from the origin of the attacks toward the targets/victims (in black, China; in red, Russia; in dark blue, Iran; in orange, North Korea)*

The work of the CSIS practically ignores the victim status of China (it does not list the cyberattacks this country may have suffered).

In this international environment, only a few states would seem to be the great predators, making use of any means, attacking a wide spectrum of targets worldwide.

But more generally, we also notice that cyberattacks seem to concentrate on specific geographical areas: the Middle East, Asia, even between bordering countries, or between countries which are already immersed in situations of tension or confrontation (India/Pakistan; North Korea/South Korea; North Korea/Japan, etc.) This could mean that the attackers have their own attack surfaces.

There is no fundamental difference between the power balances or power relations described by these statistics, the Covid period and the years preceding it. The main cyberattack countries of origin seem to be the same, as well as the "duels".

The image of the international balance of power (or power relations) reflected by these data is consistent with Western discourse, particularly that of the United States concerning the state of the threat – in general, and not just cyber – around the world. Several hypotheses can be formulated: either the CSIS approach is politically oriented, its analyses feed on the official positions of the American government and its security agencies and the data have not been selected in a neutral manner (in this respect, we cannot but stress the conspicuous absence of the United States from the main powers at the origin of cyberattacks, and the relatively small perimeter of states falling into this category, thus suggesting that the global cyber threat may be geographically concentrated), or the data are objective, but incomplete and may reflect an image in partial conformity with reality. It would be adventurous to claim to draw geopolitical lessons from the information extracted from the work of the CSIS, which is evidently fraught with multiple weaknesses: Latin America is obviously under-represented; attacks targeting or emanating from the African continent do not specifically identify any country; and again, the cyberattacks affecting China, Russia, India, but also Japan, are insufficiently dealt with. The reading is American-centric, portraying North America as a victim, and unevenly focusing on the main countries hostile to it.

1.4. Conclusion

Comparing various international scenarios, this work confronted data on cybercrime and contextual data in order to try to understand to what extent the changes experienced by societies during the Covid crisis were accompanied by specific developments in cybercrime: did the health crisis, and the resulting lifestyle changes, create the conditions for an acceleration and reconfiguration of cybercrime?

Here is a review and a summary of all the findings and main hypotheses formulated on the basis of the observations made.

– Connection time versus use transformation: Before the pandemic, citizens already spent a lot of time online. Due to the high penetration rate of mobile telephony in the world, before the crisis, citizens were already practically connected at all times. Thus, it is not so much the connection duration that evolved with lifestyle changes, but the nature of the services involved (teleworking applications, videoconferencing, online courses, as well as e-commerce, online games, etc.). Hence, the most important variable is not the "connection time", but the "online uses, applications" variable.

– Longer time spent by citizens in unsafe places in cyberspace: The nature of the place (and the time spent in that place) in which the potential victim moves is one of the key factors when considering the degree of exposure to risk. During lockdown, the daily life of citizens certainly changed location (from the outside, from the company, to the home), but it was not the fact of staying at home that particularly acted as a condition to risk exposure. It was the fact of accessing systems, networks and applications, which were poorly and badly secured or not secured at all, and of accessing parts of cyberspace which are risk sources (because they were hacked by cybercriminals), exposing ourselves to data theft by using e-commerce sites, etc. By using parts of cyberspace that are high-risk areas, Internet users increasingly exposed themselves. What differentiates practices in business from practices at home is the level of security of connections, the applications used and the prohibitions or constraints, which can be imposed in businesses and forbid access to places in the cyberspace which are at risk.

– For cybercriminals, the effort to adapt to the new context was marginal: In times of crisis, there was at least an adaptation of cybercriminal tactics. The thematization[17] of cybercriminal actions is an example of this adaptation at the minimum [CIS 20]. Existing cybercrime tools and attack methods remained effective. Cybercrime took advantage of the crisis context without the need for a radical transformation of its organization and attack tools. In 2020, the cybercriminal dynamic or logic already at work was essentially the same as during the financial crisis of 2007–2009. Cybercrime had then taken advantage of the context of both panic and uncertainty to launch vast spamming and phishing operations targeting banks and credit organizations.

---

17. A distinction should be made between the "thematization" and "politicization" of cyberattacks. Thematized cyberattacks are not necessarily politicized. During the first months of the pandemic, cybercrime exploited operations on the theme of Covid: phishing, malware, registration of domain names containing terms related to Covid-19.

At the moment, researchers insist on the existence of a correlation between market volatility and the emergence and growth of new cyber threats. These findings could suggest that cybercrime thrives in contexts of instability [MAT 11].

– No reversal of cybercrime trends in the long term: During the health crisis, the trend in the evolution of cybercrime was not reversed. Long-term trends prevailed. Cybercrime had steadily increased over the previous decades.

– No reversal in the relationship between online and offline crime: Cybercrime (online crime) did not surpass conventional (offline) crime in terms of volumes (number of cases registered, number of people arrested, tried, sentenced, number of complaints, etc.). During the crisis period, an increase in cybercrime was observed (or some of its forms) against a decrease in conventional crime (or some of its forms)[18], but in global terms, cybercrime was unable to impose itself, during the pandemic, over the more conventional forms of crime and delinquency. Some examples will suffice:

- In the United Kingdom[19], out of a total of 5,775,550 crimes recorded by the police in 2019, online crimes numbered 117,583, that is, around 2% of the total.

- In the United States, the number of computer fraud cases treated in 2019 represents approximately 0.2% of the total cases handled by the Department of Justice (i.e. 130 cases out of a total of 63,012)[20].

- In the city of Houston, the police record crimes and misdemeanors daily. Those relating to acts committed in the "cyberspace" account for approximately 0.1% of all the recorded cases (in 2020 and 2021)[21].

---

18. Available at: https://www.cbs.nl/en-gb/news/2020/10/less-traditional-crime-more-cybercrime.
19. Available at: https://www.ons.gov.uk/peoplepopulationandcommunity/crimeandjustice/bulletins/crimeinenglandandwales/previousReleases.
20. Available at: https://www.justice.gov/usao/resources/annual-statistical-reports.
21. According to the police statistics of the Houston Police, cases are notably classified depending on their place of commission ("premises"), and by "offense type". Cases falling under the "cyberspace" category were related to the following types of offenses in 2020: "intimidation", "credit card, ATM fraud", "all other offences", "impersonation", "false pretenses, swindle", etc. It does not explicitly refer to "cybercrimes", but to conventional felonious/criminal acts taking place in the cyberspace. However, "hacking/computer invasion" or "wird fraud" offense types – which specifically relate to the cyberspace – are also listed.

- In India, the National Crime Records Bureau[22] recorded 5,156,172 cases in 2019, of which 44,546 (0.86%) were cybercrimes. In 2020, out of a total of 6,601,285 crimes (IPC and SLL), 50,035 fell under cybercrime[23], that is, 0.75%. It should also be noted that crime overall increased by 28% from 2019 to 2020, while cybercrime only did so by 11%. While cybercrime continued to progress during the pandemic, it evolved less quickly than offline crime.

– Cybercriminals are teleworkers: The lockdown periods have not affected the daily lives of cybercriminals, who have been criminal "teleworkers" since the 1980s–1990s.

– The evolutions of cybercrime: Is the evolution of cybercrime the result of the impact of the Covid crisis, or are other factors at play? It is still difficult to associate the developments observed during the 2020–2022 period in terms of cybercrime with the Covid-19 health crisis alone. During this period, other events have marked the course of history: conflicts, economic crises, migration, climate, etc. The confrontation motives between the states are multiple and part of long-standing power struggles, reflected in cyberspace in the form of cyberattacks of variable extent. These actions featuring a political nature, as notably reflected by CSIS data, are added to the actions of organized and "ordinary" everyday cybercrime. However, CERT or police statistics do not make it possible to distinguish between these categories of attacks or incidents.

– The reasons for a contained increase in the effects of cybercriminal activity during the pandemic: a contained increase in cybercrime during the crisis is observed. While it is true there was an acceleration, spikes of activity during the first months of the year 2020 and spikes matching other lockdown periods (specific to each country or region concerned), the trend was altogether in line with the scenario before the crisis. The most important trend is the relatively moderate evolution of cybercrime as a whole, at least as it appears in the statistics available for the present study. While the pandemic acted as an amplifier regarding some forms of cybercrime, this was not the case for all of them. The most recent publications, addressing the effects of the pandemic on cybercrime, point to a decrease in victimization rates during the year 2021 due to the implementation of "guardianship

---

22. Available at: https://ncrb.gov.in/en/crime-in-india.
23. Available at: https://ncrb.gov.in/sites/default/files/CII%202020%20Volume%201.pdf.

measures". But the acknowledgment of cyber risks is highly contrasting, depending on the sectors of activity and countries (see section 1.6.4.). The increased use of these measures, such as the use of protective software, as well as security protocols for browsers, has contributed to this decline [SMI 22]. Behavioral changes in cybersecurity cannot be said to have given free rein to cybercrime, but may have at least limited its scope. At the same time, it would seem that cybersecurity players maintained their activities: "there has been no significant impact of the Covid-19 pandemic on the cooperation and interaction between the three communities[24] and their ability to function" [ENI 22]. But we should be cautious regarding this point because this observation cannot be generalized. A study on the impact of the epidemic on the action of the British police forces revealed that the rate of absenteeism due to medical reasons in 2020 was particularly high, resulting in a decrease in the level of satisfaction and confidence in police capabilities to deal with cyber investigations [HAL 22].

– Has cyberspace become much more "dangerous" during the pandemic? The "dangerousness" of the cyberspace had increased long before the pandemic and changed little during the health crisis (see sections 1.5.1 and 1.5.2).

– The heterogeneity of cybercrime typologies: Statistics from CERTs, police and state agencies are constructed on the basis of cybercrime typologies which differ from state to state, sometimes between actors within the same state, and are often subject to change over time within the same organization. This is translated as heterogeneity in the data series, something which complicates their analysis and the formulation of hypotheses regarding the evolution of cybercrime over long periods. The case of India in section 1.5.3 is eloquent.

– No geographical redistribution of cyberattacks during the pandemic: The geographical distribution of cyberattacks around the world seemed unaffected by the health crisis. This means that there was no substantial reorganization of the cybercriminal system, cybercrime sources remained the same and the attack surface did not depend on changes in the social or

---

24. CSIRTs, LEA, the Judiciary.

political context in the short term, but on other contingencies: technical considerations, on the one hand, (when cybersecurity progresses, when resistance increases, we can think that cybercrime moves towards other more accessible targets), and political issues, on the other hand.

1.5. Appendix

1.5.1. *Cybercrime tools: malware*

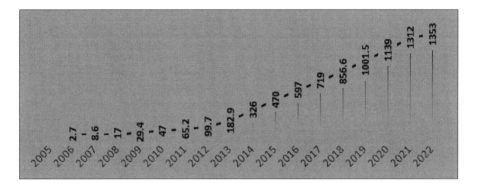

Figure 1.32. *Total number of known malware (in millions). Reconstructed from data published on AVTest: https://www.av-test.org*

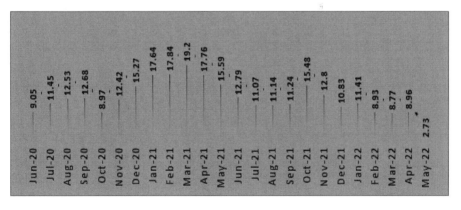

Figure 1.33. *Number of new malware software/month (in millions). Reconstructed from data published on AVTest: https://www. av-test.org/fr/statistiques/logiciels-malveillants/*

1.5.2. *CVSS as indicators of vulnerability levels*

The number of vulnerabilities detected and their level of severity is another measure of the degree of exposure to risks in the cyberspace. Data relating to the CVSS produced by the NIST illustrates the evolution of these detections and vulnerabilities around the world since 2001 (see Figure 1.34). These vulnerabilities constitute part of the exploitable attack surface and reflect the awareness level that cybersecurity actors must develop concerning weaknesses.

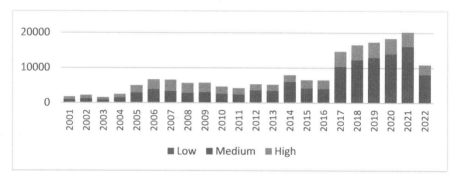

Figure 1.34. *Evolution of CVSS depending on their level of severity over time (2022 data only covers the period from January to May). Histogram reconstructed from: https://nvd.nist.gov/general/visualizations/vulnerability-visualizations/cvss-severity-distribution-over-time*

Medium-severity vulnerabilities were the most numerous over the entire period. In addition, growth was strongly accelerated in 2016. In particular, Covid years show no trend changes. Over the entire period, the number of vulnerabilities, on the three levels of severity, continued to grow.

However, the share (expressed in the following figure as a % of the annual total) occupied by each level of severity evolved differently: while the "low" severity level fluctuated around 10% of the annual total, this was not the case for the other two levels: high-severity vulnerabilities decreased from 46% in 2001 to 20% of the annual total in 2021, while on the contrary, the medium-severity level increased from 42 to 64%. Once again, the Covid years did not reverse nor did they strongly accelerate the dynamics already at work in previous years.

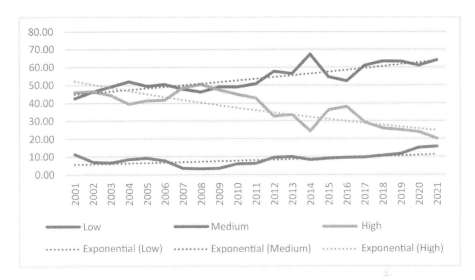

Figure 1.35. *Change in the proportion (in %) of CVSS severity levels (curves reconstructed from: https://nvd.nist.gov/general/visualizations/ vulnerability-visualizations/cvss-severity-distribution-over-time)*

1.5.3. *Heterogeneity and complexity of cybercrime typologies*

The level of detail of cybercrime typologies used in each country can be an obstacle to their international comparison. But this complexity can also act as a barrier in the process of entering information by police officers or CERT employees, or even by statistical agencies.

In India, for example, the categories of cybercrimes taken into account in the statistics of the Ministry of Home Affairs are based on the IT Act (ITA) and the IPC (Indian Penal Code). These annual statistics, whose reports have been published since 1953, added the cybercrime category in 2002. Below is a reproduction of the entire typology used by the Indian Ministry of Homeland Security in order to better illustrate this point:

– Those falling under the IT Act[25]:

  - Tampering with the computer source department.

  - Hacking:

---

25. Source: TABLE 18.6 Incidence of Cybercrimes Cases Registered During 2002 (IT ACT 2000). Available at: https://ncrb.gov.in/sites/default/files/crime_in_india_table_additional_table_chapter_reports/Table%2018.6.pdf.

a) Loss/damage to computer resource/utility.

b) Hacking.

- Obscene publication/transmission in electronic form.

- Failure:

a) of compliance/orders of certifying authority;

b) to assist to decoy or the information in interception by government agency.

- Unauthorized access/attempt to access protected computer system.

- Obtaining license or digital signature by misrepresentation/suppression of fact.

- Publishing false digital signature certificate.

- Digital fraud/signature.

- Breach of confidentiality/privacy.

- Others.

– Those falling under the CPI[26]:

- Public servant offences by/against.

- False electronic evidence.

- Destruction of electronic evidence.

- Forgery.

- Criminal breach of trust/fraud.

- Property/mark:

a) Counterfeiting.

b) Tampering.

c) Currency/stamps.

---

26. Source: TABLE 18.7 Incidence Of Cybercrimes Cases Registered During 2002 (Offences under IPC). Available at: https://ncrb.gov.in/sites/default/files/crime_in_india_table_additional_table_chapter_reports/Table%2018.7.pdf.

But the nomenclature has changed over the years. In 2020[27] (date of publication of the last volume available to the author), it is as follows[28]:

– Falling under the IT Act:

  - Tampering with the computer source department.

  - Computer-related offenses:

  a) Ransomware.

  b) Offenses other than ransomware.

  c) Dishonesty receiving stolen computer resources or communication devices.

  d) Identity theft.

  e) Cheating by personation using computer resources.

  f) Privacy violation.

  g) Cyber terrorism.

  h) Publication/transmission of obscene/sexually explicit acts in electronic form:

   i) Publishing or transmitting obscene material in electronic form.

   ii) Publishing or transmitting of material containing sexually explicit acts in electronic form.

   iii) Publishing or transmitting of material depicting children in sexually explicit acts in electronic form.

   iv) Preservation and retention of information by intermediaries.

  i) Interception, monitoring or decryption of information.

  j) Unauthorized access/attempt to access protected computer system.

  k) Abetment to commit offences.

  l) Abetment to commit offences.

---

27. Available at: https://ncrb.gov.in/en/crime-in-india-table-addtional-table-and-chapter-contents?field_date_value[value][year]=2020&field_select_table_title_of_crim_value=All&items_per_page=All.

28. Available at: https://ncrb.gov.in/sites/default/files/crime_in_india_table_additional_table_chapter_reports/TABLE%209A.2.pdf.

– Falling under the IPC (involving communication devices as medium/target or r/w IT Act):

- Abetment of suicide (online).

- Cyber stalking/bullying of women/children.

- Data theft.

- Fraud:

a) Credit card/debit card.

b) ATMs.

c) Online banking fraud.

d) OTP fraud.

e) Others.

- Cheating.

- Forgery.

- Defamation/morphing.

- Fake profile.

- Counterfeiting:

a) Currency.

b) Stamps.

- Cyber blackmailing/threatening.

- Fake news on social media.

- Other offenses.

– Reporting to SSL (involving communication devices as medium/target):

- Gambling Act (online gambling).

- Lotteries Act (online lotteries).

- Copy Right Act, 1957.

- Trade Marks Act, 1999.

These modifications affect the production of stable statistical series over long periods. In the space of two decades, the nomenclature expanded, and this phenomenon may partly explain the variations appearing in the figures (number of cybercrime cases recorded).

The nomenclature of the ministry also seems not to be used identically in the various police services across the country. For example, the cybercrime unit of the New Delhi Police uses the following typology:

– email fraud;

– social media crimes;

– mobile app-related crimes;

– business email compromise;

– data theft;

– ransomware;

– net banking/ATM fraud;

– fake calls fraud;

– insurance fraud;

– lottery scams;

– bitcoin;

– cheating scams;

– online transactions fraud.

This lack of harmonization within the same state between police services and security agencies, the diversity of typologies at an international scale and the changes made to the typologies (by adding new categories of cybercrime, for example) makes comparisons more difficult and calls for caution when interpreting the figures and statistics produced.

In Switzerland, "digital crime amounts to 33 distinct operating modes and 28 PC offenses, which are divided into five main areas:

– economic cybercrime (24 operating modes);

– cyber sexual crimes (four operating modes);

– cyber-damage to reputation and unfair practices (three operating modes);

– darknet (one operating mode);

– others (one operating mode)[29].

1.5.4. *Attitude of companies toward cyber risks: the case of the United Kingdom*

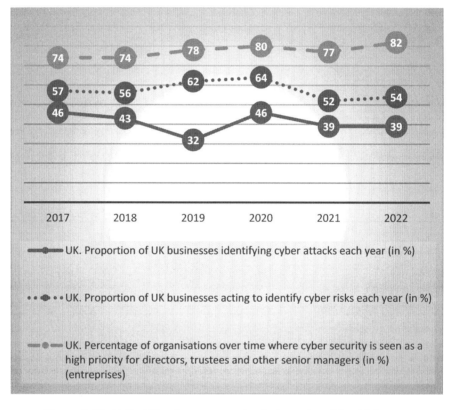

Figure 1.36. *United Kingdom, companies facing cyber risks. Histogram reconstructed from: https://www.gov.uk/government/statistics/cyber-security-breaches-survey-2022/cyber-security-breaches-survey-2022*

---

29. Available at: https://dam-api.bfs.admin.ch/hub/api/dam/assets/22164351/master.

1.6. References

[BAS 20] BASYIR M., Cybercrime in Penang shoots up 441.7 per cent since MCO, available at: https://www.nst.com.my/news/crime-courts/2020/05/594436/cybercrime-penang-shoots-4417-cent-mco, 2020.

[BUI 20] BUIL-GIL D., MONEVA A., KEMP S. et al., "Recorded cybercrime and fraud trends in UK during Covid-19", *Statistical Bulletin on Crime and Covid-19*, no. 6, p. 2, 2020.

[CHE 21] CHEN P., KURLAND J., PIQUERO A.R. et al., "Measuring the impact of the Covid-19 lockdown on crime in a medium-sized city in China", *Journal of Experimental Criminology*, pp. 1–28, 2021.

[CIS 20] CISA, Covid-19 exploited by malicious cyber actors, available at: https://www.cisa.gov/uscert/ncas/alerts/aa20-099a, 2020.

[ENI 22] ENISA, 2021 Report on CSIRT – LE Cooperation, A study of the roles and synergies among sixteen selected EU/EEA Member States, available at: https://www.enisa.europa.eu/publications/2021-report-on-csirt-law-enforcement-cooperation/@@download/fullReport, 2022.

[FIN 20] FINKLEA K., Covid-19: Cybercrime opportunities and law enforcement response, Report IN11257, CRS, 2020.

[GRY 21] GRYSZCZYNSKA A., "The impact of the Covid-19 pandemic on cybercrime", *Bulletin of the Polish Academy of Sciences Technical Sciences*, vol. 69, no. 4, doi: 10.24425/bpasts.2021.137933, 2021.

[HAL 21] HALE T., ANGRIST N., GOLDSZMIDT R. et al., "A global panel database of pandemic policies (Oxford Covid-19 Government Response Tracker)", *Nature Human Behaviour*, doi: 10.1038/s41562-021-01079-8, 2021.

[HAL 22] HALFORD E., "An exploration of the impact of Covid-19 on police demand, capacity and capability", *Social Sciences*, vol. 11, no. 7, p. 305, 2022.

[HAW 20] HAWDON J., PARTI K., DEARDEN T.E., "Cybercrime in America amid Covid-19: The initial results from a natural experiment", *American Journal of Criminal Justice*, doi: 10.1007/s12103-020-09534-4, 2020.

[HOR 21] HORGAN S., COLLIER B., JONES R. et al., "Re-territorialising the policing of cybercrime in the post-Covid-19 era: Towards a new vision of local democratic cyber policing", *Journal of Criminal Psychology*, vol. 11, no. 3, pp. 222–239, 2021.

[INT 20] INTERPOL, *Cybercrime: Covid-19 Impact*, Interpol, France, 2020.

[KAS 20] KASHIF M., AZIZ-UR-REHMAN, JAVED M.K. et al., "A surge in cyber-crime during Covid-19", *Indonesian Journal of Social and Environmental Issues*, vol. 1, no. 2, pp. 48–52, 2020.

[MAH 20] MAHADEVAN P., Cybercrime: Threats during the Covid-19 pandemic, Global Initiative against Transnational Organized Crime, Geneva, 2020.

[MAT 11] MATARNEH B., SHEHADEH H.K., "World financial crisis and cybercrime", *Int. J. Buss. Mgt. Eco. Res.*, vol. 2, no. 1, pp. 124–130, 2011.

[SÉN 20] SÉNAT, Désinformation, cyberattaques et cybermalvaillance : l'autre guerre du Covid-19, Executive summary, Paris, DLC No. 125, 2020.

[SMI 22] SMITH T., "Assessing the effects of Covid-19 on online routine activities and cybercrime: A snapshot of the effect of sheltering in place", *Caribbean Journal of Multidisciplinary Studies*, vol. 1, no. 1, pp. 36–60, 2022.

[UNO 20] UNODC (UNITED NATIONS OFFICE ON DRUGS AND CRIME), *Cybercrime and Covid-19: Risks and Responses*, UNODC, Vienna, 2020.

[WAH 21] WAHEEDA F., "New trends in cybercrime in the Maldives – Moving beyond legal measures in the new norm", *International Journal of Law, Government and Communication*, vol. 6, no. 24, pp. 104–115, 2021.

[WIG 20] WIGGEN J., *The Impact of Covid-19 on Cybercrime and State-Sponsored Cyber Activities*, Konrad Adenauer Siftung, Berlin, 2020.

2

The SARS-CoV-2 Pandemic Crisis and the Evolution of Cybercrime in the United States and Canada

2.1. Introduction

The appearance and transmission of the SARS-CoV-2 virus causing the outbreak of Covid-19 in China and around the world between 2019 and 2022 has transformed living, working and interpersonal relationship habits. This pandemic has even been described as a great accelerator of social and digital transformations [PET 21]. The sanitary measures imposed by governments and the first lockdowns, which can be described as sudden, affected billions of people worldwide [HOS 22]. This crisis generated unprecedented social transformations, among which lockdowns and social distancing policies were the most striking examples. Many societies were affected by a widespread and forced shift toward teleworking, online exchanges and dependence on remote government and healthcare services, which resulted in a significant increase in unemployment. The shift of economic activities from the offices in the cities to private residences and the poorly protected networks of individuals further exposed millions of individuals to cyberspace risks and threats [LAL 21].

In a report published in August 2020, Interpol warned:

Chapter written by Hugo LOISEAU.

With organizations and businesses rapidly deploying remote systems and networks to support staff working from home, criminals are also taking advantage of increased security vulnerabilities to steal data, generate profits and cause disruption [INT 20b, p. 4].

These major social changes triggered by the pandemic raise questions about their impact on cybercrime. Were the nature and frequency of cybercrime influenced by the shift of daily activities to cyberspace? Did cybercriminals know how to exploit the opportunity offered to them to make their criminal activities thrive and, if so, how did they proceed? This chapter will study the impact of the SARS-CoV-2 pandemic on cybercrime in the United States and Canada, by exploring the following general research question: how did the SARS-CoV-2 pandemic and its consequences affect cybercrime in the United States and Canada?

The first part of this chapter will draw a rapid portrait of the pandemic crisis in Canada and the United States and discuss societal changes, namely the acceleration of society cyberization. Targets, victims and malicious actors (cybercriminals, in other words) will be introduced in the second part of the chapter, in the form of a literature review. Then, the impacts of the health crisis on the evolution of cybercrime in Canada and the United States will be explored. In the fourth part, the data presented in the previous part will be analyzed in depth, to finally conclude on the difficulties encountered throughout the study.

2.2. The impacts of the SARS-CoV-2 pandemic

The SARS-CoV-2 pandemic was a health crisis engendering significant changes in society, activated by the way in which governments reacted to it. The compulsive shift of workers toward teleworking [CAM 22], the reduction of mobility [HOS 22], the shift toward remote teaching in education [TAD 20] and the use of the Internet for social contacts restricted by bans on social meetings accelerated the cyberization of entire sections of the world's population. Not only did the pandemic crisis alter the evolution of crime [REG 22], but also that of cybercrime.

On a strictly epidemic level, as demonstrated by the two graphs (see Figures 2.1 and 2.2) compiled by the World Health Organization (WHO) since the start of the pandemic in 2020, the SARS-CoV pandemic-2 affected Canada and the United States in a different manner. The number of victims and the prevalence of Covid-19 largely differed between the two states.

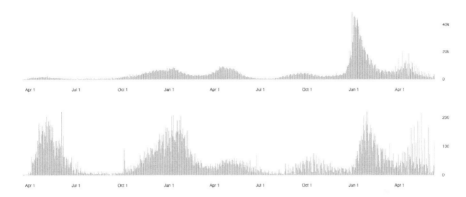

Figure 2.1. *Confirmed cases of Covid-19 in Canada [WHO 22a]*

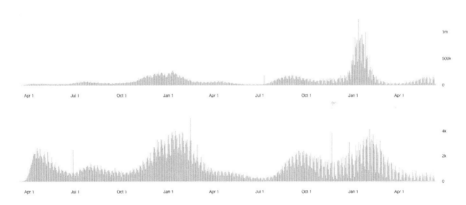

Figure 2.2. *United States of America's situation [WHO 22b]*

The impacts of the pandemic and the ensuing crisis followed a clearly defined timeline in North America. According to the *American Journal of Managed Care*, on January 9, 2020, the WHO publicly exposed the case of pneumonia associated with a new coronavirus in the city of Wuhan in China [AJM 21]. On January 31, 2020, the WHO declared a global health emergency, with a striking number of cases over 9,800 worldwide. The first case detected in the United States was announced on January 21, 2020, and on January 25, 2020 in Canada. On March 11, 2020, the WHO declared Covid-19 a pandemic [AJM 21].

On March 13, 2020, a federal health emergency was declared in the United States. The federated states distinctly declared it between March 1 and March 15, 2020 at the latest [AJM 21]. In Canada, Quebec declared a provincial state of emergency on March 13, Ontario and Alberta on March 17, British Columbia and Saskatchewan on March 18, New Brunswick on March 19 and Manitoba on March 20, 2020 [CAN 21].

The consequences of this pandemic and the government responses overturned the lifestyles of Canadians and Americans. According to a study conducted by the Canadian Internet Registration Authority (CIRA) comparing the data of their Canada's Internet Factbook, between 2019 and 2020, during the lockdowns, 54% of Canadians worked from home in a teleworking format, whereas in 2019, less than 42% had occasionally worked from home [AUT 20]. In the United States, during the month of May 2020, 35.4% of the population aged 16 and over worked from home due to stay-at-home orders [USB 20].

2.3. Cybercrime and SARS-CoV-2

During this health crisis, cyber risks and cyber threats increased [KAT 22]. A cyber risk is described as the product of the level of threat and the level of vulnerability. While cyber risks determine the likelihood of a successful cyberattack [SAN 22], cyber threats represent the potential for violation of security, which exists when there are circumstances, capabilities, actions or events that could breach security and cause harm. Grasping these two notions is fundamental to understand, because the cyberization that societies have experienced over the last 20 years has largely contributed to the complexity of cybersecurity issues. Back in 2001, Beck already heralded that "the social production of wealth is accompanied by the social production of risks" [BEC 01, p. 36]. The SARS-CoV-2 pandemic in 2020 and the ensuing multidimensional crisis (social, economic, political crisis) represent a contextual window which perfectly illustrates the risk society Beck refers to, characterized by the multiplication and diffusion of systemic and cross-sectoral risks stemming from technological and industrial developments. Globalization and cyberspaces have a multiplier effect on these risks and the occurrence of crises [BEC 01]. During this period, crime in general and cybercrime in particular have developed because of the risk society, globalization and cyberization [SAL 22].

The 2020 crisis context provoked by the SARS-CoV-2 pandemic attests to this, since risks grew hand in hand with vulnerabilities. New social interactions, caught within the offline–online continuum, brought to light the mutual impacts between tangible reality and cyberspace [WAL 17], manifest in all societies fighting the pandemic. Indeed, this period of crisis led to an acceleration of the cyberization movement born in North America around 1995. The targets and victims of cybercrime shifted many of their activities to cyberspace, be them individuals or public and private organizations. For some authors such as Hawdon et al., the pandemic crisis was able to illustrate the criminological routine activity theory [COH 79] in a natural context. Stricto sensu, according to these authors, the societal changes forced by the stay-at-home orders quantitatively and qualitatively increased cyber risks. This occurred because three conditions were met in the same space-time, namely a pool of potential victims combined with the presence of motivated malicious actors and the lack of an effective guardian [HAW 20].

Out of opportunism, malicious actors adapted and intensified their techniques as well as their attack vectors following the crisis context, whereas other types of crimes decreased [MOH 20]. Paradoxically, according to Katagiri, the state of cyberspace stability was more durable than initially thought, in the face of sudden transformations in the behavior of Internet users and organizations. According to this author, "hardly any substantive change has occurred to the structure of cyberspace dynamics, power relations, and strategic interests of major cyber powers" [KAT 22, p. 1], which demonstrates the structural tenacity of cyber anarchy, even in times of crisis. However, the number and nature of the targets and potential victims changed considerably.

2.3.1. *Targets and victims*

The shift of workers from a labor environment controlled and protected by implanted and monitored systems to private residences involves risks which can increase the vulnerability of businesses and workers to cyberattacks. According to SentryBay, a cybersecurity company, in 2020, 42% of residential and private terminals were not sufficiently protected in the United States [SEN 20]. These terminals, unprotected or poorly protected, represented a possible gateway for malicious actors wishing to attack networks and damage business data. "With so many people using compromised laptops or home computers to log-in to the corporate network, they are creating a weak link in the security chain, and potentially devastating damage to their employer at what is already a very testing time" [SEN 20].

Since the pandemic started, the Federal Bureau of Investigation (FBI) in the United States warned private organizations as to the need to protect their networks in the era of teleworking. According to their warning notice, cybercriminals frequently carry out attacks on the private terminals of teleworkers to inject malicious software and code into the organization's protected system, to commit fraud and to gain deeper access to the network of the company and its employees [FBI 20].

As with telework, educational institutions were forced to close during the first wave lockdown. According to UNESCO, more than 1.6 billion pupils and students in approximately 165 states were affected by preventive health measures [UNE 20]. These figures represent approximately 87% of all the pupils and students worldwide. In several cases, schools and higher education institutions were forced to close either for the entire remaining term (Winter 2020) or until the end of the standard school calendar [LAW 22, WIK 22a, WIK 22b]. Teaching, at first temporarily halted for a period ranging from a few days to a few weeks, completely moved online. Remote teaching and videoconferencing services were required by school services. The choice and implementation of these services were disparate, depending on regions and socioeconomic characteristics [MOR 20]. On this subject, security issues were raised by several experts regarding the choice of services and their implementation method [BAN 20]. As was the case with teleworking, security issues were related to user habits and cyber hygiene. Unsecured student terminals could potentially open the door to cyberattacks against systems connecting hundreds of thousands of users. Moreover, given the audience of online teaching tools, some authors warned against the safety risks of minors using services which required the use of webcams and file transfer [OLO 20, COL 20].

The closure of several businesses and physical service points as a result of the lockdown caused a shift of customers to services and online sales sites [FBI 20]. According to the report by Absolunet, a consulting service company, certain sectors such as "furniture and home furnishing", "food and catering", "clothing", "sports and leisure goods" and "household appliances, electronics, materials construction and renovation" witnessed an online sales growth of around 106%, 160%, 21%, 105% and 161% respectively, compared to 2019 [ABS 20].

However, according to Hawdon et al., online shopping continues to be one of the online routines deemed the most dangerous [HAW 20]. For reference, these same authors also consider the use of social media, online gambling and visiting the dark web as practices increasing user risks and of systems being targeted by cyberattacks or malicious actors. An increase in online transactions implies that personal data are more vulnerable than when doing in-person purchases at a shop. These data were targeted by cybercriminals wishing to commit personal data theft, identity theft or fraud [FBI 20].

It must be noted that relations with cyberspace deepened during the different phases of the pandemic crisis in North America. A shift from physical activities to online activities was observed in many areas. These rapid and forced changes to cyberspace had significant consequences on social relations, government decision-making and on cybercrime. In this regard, both individuals and organizations were affected by cybercrime.

With societal changes and the obligation for many individuals to change their living, working, socializing and entertainment habits, several authors have argued that lockdown measures have actually increased cybercrime practices targeting individuals. According to Ahmad et al., cybercriminals wishing to attack individuals used social engineering to exploit fears and widespread anxiety around the topic of the pandemic [AHM 20, ALZ 20, LAL 20]. This has been defined by Ventrella as "the science of using social interactions" [VEN 20, p. 389]. Cybercriminals became closely interested in individuals and tried to exploit the new vulnerabilities having emerged with the pandemic crisis. Alzahrani has described several types of attempts exploiting the fear, hope or lack of cyber hygiene of individuals: malicious attachments, phishing campaigns focusing on physical or financial health, fake online sales websites allegedly selling protection products, extortion or theft of conversations, as well as brand impersonation [ALZ 20]. According to these authors, the purpose is to convince individuals that they are dealing with the material or sites by legitimate and existing organizations [VEN 20, OLO 20].

This type of attack is aimed at a large number of targets in order to increase the probability of success. Although the message has to be targeted, it also needs to be broad enough so as to address a large audience, such as claiming to have a cure or offering financial help in a context of distress, at a moment when unemployment rates and job insecurity are particularly acute.

Despite the fact that many of these attempts are made with the intention of stealing, directly victimizing individuals or defrauding government financial assistance programs, some malicious actors target workers, and then access corporate networks or organizational security. Individuals rather became a gateway to the real victim, which actually was the targeted organization [ALZ 20, MIC 20].

Children, whose daily lives also shifted toward cyberspace during the first wave of the pandemic, were other possible victims of cybercrimes targeting individuals. According to the FBI, the closure of schools forced distance learning at the time it paved the way for the victimization of children via online education platforms. The FBI's newsletter Private Industry Notification stated that platforms provide broader "access to cybercriminals to collect data and monitor children, potentially providing opportunities to track and attack minors" [FBI 20]. These findings were also shared by Europol and Olofinbiyi [EUR 20, OLO 20]. According to them, the greater presence of children using electronic devices increases the risks of being exploited by abusers or traffickers, together with the sexual exploitation of children online.

In the specialized literature, individuals are presented both as victims targeted by certain cybercriminals wishing to implement rather simple attacks (while having the possibility of producing as many victims to raise as much money as possible), and as the access doors to better protected networks which would not be so easily accessible to cybercriminals under normal circumstances.

Organizations, businesses and government agencies were also victims of an increase in cybercrime during the pandemic. According to Ventrella, just as individuals suffered, organizations were put under pressure by the social engineering of cybercriminals [VEN 20]. However, according to the FBI, these were mainly brute force attacks, committed directly [FBI 20]. According to Ventrella, while personal data theft or identity theft were the most common offenses against individuals, in the case of companies and organizations, ransomware was the type of attack that was most often observed during the pandemic [VEN 20].

Some reports and authors also included healthcare services as victims of choice for cybercriminals [CEN 22]. Hospitals were targeted by denial-of-service attacks in exchange for a ransom. Criminals attacked these organizations probably expecting them to pay ransoms more quickly, given the urgency and the challenge of finding an IT service in times of pandemic, so as to be able to protect patients and continue to offer health services.

2.3.2. Malicious actors

While the victims of cybercrime are known, it is more difficult to accurately identify the malicious actors behind this phenomenon. Part of the literature on cybercrime – and without being limited to it – is dedicated to the more general study of major crime. To this end, Gayraud proposes four prominent characteristics of major crime. (1) Major crime chiefly manifests itself by polycriminality. Major criminal groups are opportunistic and pragmatic in criminal markets, in the sense that they do not necessarily develop a specialization of their criminal practices. (2) These groups are territorialized, rooted in a space that allows them to create their own biotope, setting up hermetic enclaves inaccessible to the public authorities. This favors their territorialized and immaterial expansion in the cyberspace. (3) These groups and organizations are considered unsinkable. This is because they are highly adaptable to socioeconomic changes and resistant to repression by the public authorities or to the competition from other criminal groups. (4) Finally, these groups have major macroeconomic impacts since they manage massive, globalized and interconnected financial flows that facilitate and promote corruption and the laundering of their illicit income [GAY 21]. These four characteristics partly explain the rise in cybercrime noticed around the world as indicated in the 2020 report from the United Nations Office on Drugs and Crime (UNODC) on the Covid-19 pandemic and organized crime:

> Nevertheless, the disruption caused by Covid-19 has been quickly exploited, with some criminal groups expanding their portfolio, particularly in relation to cybercrime and opportunistic criminal activities in the health sector [UNO 20, p. 11].

Moreover, cybercrime and its study do not only concern organized crime, but also daily, even ordinary crime, which greatly vary depending on socioeconomic backgrounds and world regions [REG 22]. For example, it is possible to mention online defamation, violent extremism and hate speech on the Internet and social media [BEN 22], petty crime [COL 20], radicalization [ALA 21], disinformation [PAR 20], etc. which are common forms of cybercrime.

Major criminal groups and ordinary cybercrime play a major role in what might be termed as a "criminological diffuse background" whose omnipresence and pervasion in the cyberspace act as the basis for many illicit practices of interest to cybersecurity [BRE 10]. For example, the perpetration of a cybercrime, when left unpunished, is the preliminary stage for cybercrime, cyber espionage and cyberattacks in general[1]. As a result, cybercrime benefits from an immense market bringing together the supply (of software, services and available techniques in cyberspace) and the demand from criminal or non-criminal organizations, states and individuals whose goal is to exchange a good (data) by means of an increasingly dematerialized and fungible currency: cryptocurrency [BAD 20].

2.3.3. *Cyberspace: a propitious environment for cybercrime*

The concept of cybercrime is key to this research. Here, it is the definition reformulated by Buil-Gil et al. [BUI 20] that will be used, which considers cybercrime as a set of crimes including cybernetic crimes and cyber-dependent crimes. According to these authors, cybercrimes are crimes made possible solely by the existence of computer systems and networks. These crimes include, for example, hacking, production, distribution and use of computer viruses, and denial-of-service attacks. Cyber-dependent crimes are those existing in a traditional form, but having increased in scope and number because of cyberspace. These crimes include, for example, online fraud and phishing.

For the Royal Canadian Mounted Police (RCMP), cybercrimes are divided into two categories. In operational terms, this distinction facilitates the fight against cybercrime. These two categories rejoin and include the

---

1. The perpetration of a cybercrime violates the principles of information system security, namely the confidentiality, integrity and availability of data. These principles are qualified as crimes in the vast majority of penal or criminal codes around the world.

categories proposed by Buil-Gil et al. On the one hand, there are the offenses where technology is the target and, on the other hand, there are those in which technology is the instrument [GRC 14]. Figure 2.3 illustrates this definition.

Figure 2.3. *Cybercrime categories by the Royal Canadian Mounted Police [GRC 14]*

The illegal nature of the actions deployed by malicious actors forces them to camouflage their actions. To this end, cyberspace offers protection and a propitious environment to carry out their activities. The anonymity [LEC 10] and the obfuscation that cyberspace enables [KOT 17] the low cost of cybercrime compared to the potential gains [DEC 17] and the lack of expertise of police and judicial bodies [BAU 14, KAI 21] are detrimental to the fight against cybercrime. Furthermore, the fluid (data being digital), plastic and fungible nature of data (because of cryptocurrencies) greatly simplifies the perpetration of cybercrimes.

Cyberspace brings together a variety of essential layers which work as an interconnected, functional network [DUP 17]. Although they constitute the functional body of cyberspace, the physical, software and informational layers elicit little interest on the part of the political field. On the contrary, it is the social layer which is particularly relevant for (and arouses the interest of) politicians and political scientists. This stratum includes both individual interactive behaviors with the cyberspace and a collective component weighing on policies, institutions, laws, and standards, regulating and framing cyberspace interactions and use. Despite the fact that cybercriminals may be interested in flaws in software or physical systems to commit their crimes, the social character is also fraught with vulnerabilities which can be perceived and exploited by cybercriminals. Given the fact that each Internet user is free to act in cyberspace and that cyber hygiene is not always appropriate and sufficient, users are both possible victims and vectors, functioning as system gateways. Systems are more vulnerable from the very moment a user behaves unsafely.

Moreover, in virtue of its supranational nature, cyberspace is difficult to govern and secure. The governance of the Internet and cyberspace is suboptimal [KAT 21] and polycentric [DUP 16], which limits the fight against cybercrime, revealing one of the major political flaws in cybersecurity. Even in developed states, cybersecurity management is referred to as a confusing patchwork [CHA 18]. Given the amount of information passing through the networks at any time, supervising and controlling information and transactions, as well as verifying the legitimacy and legality of content is almost impossible for the authorities. In this way, supervisory, governmental, organizational or corporate bodies may take too long to process information and denunciations. This can affect the detection and recovery phases, making the systems inoperative.

The safety and protection of networks are considered to be a shared responsibility among security and law enforcement agencies, government bodies, businesses, organizations and individuals. In recent years, several social changes have occurred, characterized by an increase in the dependence on cyberspace. Policies and monitoring techniques do not always follow the speed of the advancements made in this area. This is especially true in the event of a crisis such as the one experienced by all countries worldwide during the SARS-CoV-2 pandemic in 2020.

Technical cybersecurity vulnerabilities are well known. In addition to these, individual and social human vulnerabilities expand the universe of possibilities available to malicious actors in cyberspace. Given the recent nature of the events analyzed – and despite the general observation in the scientific literature that cybercrime seems to have increased during the pandemic crisis – little scientific research is currently available regarding the specific cases of Canada and the United States. This is why the research question of this chapter inquires whether the pandemic crisis has influenced the evolution of cybercrime in these two countries. Did malicious actors adapt their practices to the new social context created by the pandemic crisis and government measures?

The present study of Canadian and American cases draws on various sources (research reports, statistics, the opinion of police and surveillance bodies, public and private organizations) having coped with cybercrime during the pandemic crisis. The purpose of this chapter, though, is not to describe and analyze all the cybercrimes reported during the pandemic, but rather to provide an overview of the contrasting evolution of cybercrime in Canada and the United States, and its impact on individuals as well as on private and governmental organizations.

2.4. The evolution of cybercrime in North America during the pandemic

Among all the government measures taken to combat the pandemic and its impacts, lockdown was one of the most significant and effective measures. Significant in terms of the restrictions on freedom of movement in order to limit the circulation and retransmission of the SARS-CoV-2 virus, this upheaval led to a shift of activities to the virtual space, notably, teleworking. Cyberization has increased suddenly and sharply, transferring work and study from home to digital spaces lacking cyber-adapted and sufficient safeguard to enable user protection from virtual threats.

For example, in April 2020, in Canada, 67% of people with a higher education degree were working from home, compared to only 8% in 2018. For people whose highest educational level is secondary school or less, this rate rose to 16% over the same period, compared to 2% two years earlier [TAH 21]. As previously stated, the diversity and extent of lockdown experiences in Canada and the United States being highly contrasting, it seems of little use to extensively describe them in this chapter. The purpose of this chapter is rather to describe the impact that the pandemic and the different lockdown experiences had on cybercrime in North America.

Many reports published during the pandemic by government security organizations discuss information and awareness among digital users against the dangers and threats developing in this dematerialized space. Studies aimed at teleworkers, companies, health services or any other potential victims of malicious actors in cyberspace are abundant. The majority of resources available and freely accessible on this subject discuss a variety of topics, ranging from protection against ransomware to spotting Internet fraud. This type of communication does not necessarily provide quantitative information about cyber risks, but only strategies to prevent them.

2.4.1. *The United States*

In the United States, the fight against cybercrime falls within the jurisdiction of the federal government. Several law enforcement agencies are in charge of cybersecurity in general, and the fight against cybercrime. To carry out their mission, this fight is subdivided into numerous units or task forces dedicated to fighting against specific aspects of cybercrime (fraud, forensics, child exploitation, protection of critical infrastructures, etc.). Apart from the Department of Homeland Security, the Department of Justice and the Cybersecurity & Infrastructure Security Agency (CISA), the FBI is at the front line of the fight because of the Internet Crime Complaint Center (IC3, equivalent to a CERT), created in May 2000.

The IC3 publishes reports keeping log of the cybercrimes reported by individuals or businesses every year. The main interest of the information delivered to us through these reports lies in the comparison of the number of complaints filed each year to be able to establish broad trends or inflections in cybercrimes. However, the information provided by IC3 reports does not suffice to grasp the extent of cyber threat phenomena: it is necessary for a complaint to be filed for the wrongdoing to be recorded and computed [INT 20]. It should be noted that the 2021 IC3 report no longer kept track of certain types of crimes (such as "charity fraud"), merged other crime types into a single category (e.g. "terrorism" and "harassment/threats of violence" were grouped under "terrorism/threats of violence", in 2021), and created new crime categories ("civil matters") for their statistics. This is why only reports from 2018 to 2020 were taken into account in Tables 2.1, 2.2 and 2.3 presented below.

The IC3 2020 report clearly states that cybercrime increased and that malicious actors took advantage of the crisis caused by the pandemic:

The global impact was unlike anything seen in recent history, and the virus permeated all aspects of life. Fraudsters took the opportunity to exploit the pandemic to target both business and individuals. In 2020, the IC3 received over 28,500 complaints related to Covid-19 [INT 20, p. 9].

Then follows a series of cybercrime examples (perpetrated and reported), as well as statistics illustrating the extent of the phenomenon.

According to the 2021 version of the report, the total crime-related economic losses amounted to 6.9 billion USD, whereas 2020 recorded losses for "only" 4.2 billion USD. This amount has been constantly increasing in recent years: it was 1.4 billion USD in 2017, then 2.7 billion USD in 2018 and 3.5 billion USD in 2019 [INT 21]. Arguably, these numbers are underrepresented, especially considering that many malicious actors targeted government agencies during the pandemic crisis, such crimes not being computed in this report.

The crimes the most often reported to the IC3 center over five consecutive years are eloquent. On top of the crime list in 2021 were phishing and other similar crimes, as well as its different enforcement techniques (by telephone call, by instant messaging, etc.), identity theft, violation of personal data, non-payment or non-delivery and extortion [INT 20, INT 21]. In the case of extortion, while the number of complaints undoubtedly increased, the economic losses of victims decreased from over 107 million USD in 2019 to almost 71 million USD in 2020. In all other cases, the complaints increased along with victim losses [INT 20]. Interpol offers the same diagnosis on the basis of different sources [INT 20b].

It can be noticed that at least three other types of complaints increased alarmingly, among which crimes against children more than doubled, attacks on investments also doubled and identity theft multiplied by five.

Certain increases in crimes are also remarkable in that they first tended to decrease between 2018 and 2019, and then increased again in 2020, sometimes even exceeding the number of complaints reported in 2018. Examples include charity fraud, crimes against children, credit card fraud, harassment or threats of violence and technical support fraud [INT 20].

	2020	2019	2018	Increase from 2018 to 2020 (percentages rounded to hundredths)
Phishing/vishing/ smishing/pharming	241,342	114,702	26,379	+815%
Non-payment/ non-delivery	108,869	61,832	65,116	+67%
Extortion	76,741	43,101	51,146	+50%
Personal data breach	45,330	38,218	50,642	−10%
Identity theft	43,330	16,053	16,128	+169%
Spoofing	28,218	25,789	15,569	+81%
Misrepresentation	24,276	5,975	5,959	+305%
Confidence fraud/romance	23,751	19,473	18,493	+28%
Harassment/threats of violence	20,604	15,502	18,415	+12%
Credit card fraud	17,614	14,378	15,210	+16%
Employment	16,879	14,493	14,979	+130%
Tech support	15,421	13,633	14,408	+7%
Real estate/rental	13,658	11,677	11,300	+21%
Investment	8,788	3,999	3,693	+138%
Crimes against children	3,202	1,312	1,394	+130%
(Telephony) Denial of service/TDoS	2018	1,353	1,799	+129%
Healthcare related	1,383	657	337	+310%
Charity	659	407	493	+34%

Table 2.1. *Number of increasing cybercrimes reported to IC3 from 2018 to 2020 [INT 20]*

Among the crimes on the rise, most are related to online economic activities (such as investing, real estate and rental, credit card fraud), identity theft or fraud (confidence fraud, misrepresentation), threat or breach of trust (harassment/threats of violence, crimes against children, extortion). Many of the cybercrimes committed during the pandemic and discussed here can be explained by the migration of a large part of the population's activities to cyberspace, this population not being necessarily educated or properly informed about such space and thus representing an easy target [INT 20].

Among the relatively stable cybercrimes is ransomware. However, this information should be put into perspective with the economic losses related to this type of attack. While the number of attacks steadily increased, the financial damage rose from around 9 million USD in 2019, to over 29 million USD in 2020, more than tripling in one year [INT 20].

The cyber threat related to ransomware did not diminish, but mutated during the pandemic. Hackers have shifted their focus to targets willing to pay more money, rather than increasing the number of attacks. During the pandemic, this phenomenon resulted in the migration of attacks from individuals and businesses to government or health services, targets generally willing to pay more money and more quickly because they cannot afford to have inactive services for too long.

	2020	2019	2018	Increase from 2018 to 2020 (percentages rounded to hundredths)
Overpayment	10,372	10,842	10,826	−4%
Lottery/sweepstakes/ inheritance	8,501	7,767	7,146	+19%
Intellectual property rights/copyright and counterfeit	4,213	3,892	2,249	+87%
Ransomware	2,474	2,047	1,493	+66%
Gambling	391	262	181	+116%
Terrorism	65	61	120	−46%
Hacktivist	52	39	77	−32%

Table 2.2. *Number of stable cybercrimes reported to IC3 from 2018 to 2020 [INT 20]*

Interpol's 2020 report on cybercrimes committed during the pandemic crisis particularly warns against malware, identified as a major threat [INT 20b]. However, the number of complaints reported by the FBI in 2020 tended to decrease in this category. This can be explained by the fact that the organization IC3 reported complaints filed by individuals and organizations located in the United States. However, the same Interpol report indicated that ransomware attacks, initially directed toward individuals and businesses, had been redirected toward government agencies as well as the health sector, which are not computed according to the IC3 method [INT 20b].

On a similar note, compromise attacks and re-shipments declined, although economic damage respectively rose from 1.7 and 1.7 million USD in 2019 to 1.8 and 3 million USD in 2020 [INT 20].

	2020	2019	2018	Evolution from 2018 to 2020 (percentages rounded to hundredths)	Decrease from 2019 to 2020 (percentages rounded to hundredths)
Business email compromise (BEC)/ Email account compromise (EAC)	19,369	23,775	20,373	−5	−19%
Advance fees	13,020	14,607	16,362	−22	−11%
Government impersonation	12,827	13,873	10,978	+17	−8%
Malware/ scareware/ virus	1,423	2,373	2,811	−34	−40%
Re-shipping	883	929	907	−5	−5%

Table 2.3. *Decreasing number of cybercrimes reported to IC3 from 2018 to 2020 [INT 20]*

The IBM company in their cybersecurity body's report entitled *X-Force Threat Intelligence Index* of 2021 [IBM 21] outlined the proportions of each type of cyberattack in the United States[2]. Based on network surveys and requests from private companies, IBM was able to provide insight into the evolution of cyberattacks and cybercrime in 2020. Ransomware marked a real difference for the year 2020 compared to 2019, with a 20% increase and topping the list of the 20 most often observed cyberattacks in the United States (33% of the total).

---

2. IBM refers to cyberattacks and cybersecurity incidents with no difference in their report, as mentioned on page 7: "attacks" and "incidents" are used interchangeably in this report. An incident refers to an organization's hotline call to the X-Force Incident Response team that results in the investigation and/or remediation of an attack or suspected attack". IBM Security, X-ForceThreat Intelligence Index 2021. Available at: https://www.cert.hu/sites/default/files/xforce_threat_intelligence_index_2021_90037390usen.pdf.

Type of cyberattack	%
Ransomware	33%
Business email compromise	14%
Data theft and leakage	12%
Internal offenses	8%
Server access	6%
Web-scripting	4%
Spam	4%
Remote administration tool (RAT)	4%
Wrong configuration	2%
DDoS	2%
Credential theft	2%
Trojans to banks	2%

Table 2.4. *Proportion of each type of cyberattack in the United States in 2020 [IBM 21]*

The pandemic also made three industrial sectors particularly vulnerable (manufacturing, professional services and wholesale trade) because these sectors are sensitive to downtime in their online operations. In addition, the report claims that the healthcare sector was especially targeted in 2020, which can be seen in: "[…] the heavy targeting that healthcare received during the Covid-19 pandemic in 2020, from ransomware attacks to threat actors targeting Covid-19 related research and treatments" [IBM 21, p. 43].

On the other hand, lockdown measures reduced the number of cyberattacks targeting the transport sector. As the sector was less profitable due to the lockdown, there was an activity shift and adaptation on the part of malicious actors during the pandemic [IBM 21, p. 45].

2.4.2. Canada

In Canada, the Royal Canadian Mounted Police's (RCMP) National Cybercrime Coordination Unit (NCU), created to combat cybercrime, officially began operations on April 1, 2020. In the very words of its director, Mr. Chris Lyam, "At the time Covid hit Canada at the beginning of the year [2020], the group was not even in operation yet […] [NOR 20]" which forced the new organization to rely on the Canadian Anti-Fraud Center to accomplish its mission. This coincidence between the start of the pandemic and the creation of the NC3 was partly due to the slowness of Canadian political and police authorities to take cybercrime issues into consideration. Indeed, the first RCMP report on cybercrime dates back to 2014 [GRC 14] and it precedes the first (and only cybercrime strategy to date) in 2015 [GRC 15], 15 years after the creation of the IC3 in the United States.

Through the Canadian Center for Cybersecurity (CCC[3]), the Canadian government published its "National Cyber Threat Assessment 2020", where the impact of the pandemic crisis on cybersecurity in Canada is clearly referred to: "[…] as devices, information, and activities shift to the Internet, they also become vulnerable to cyber threat actors" [CCC 20, p. 5]. In other words, the vulnerability of systems, information and populations was at higher risk because the pandemic gave new opportunities to cybercriminals.

The authors of the report noted an increase in the number of perpetrators of cyber threats and that these threats were becoming so sophisticated that, of all the cyber threats identified, it was cybercrime that affected organizations and individuals the most. However, and the nuance here is important, while cybercrime was the most likely of cyber threats, it was not the most dangerous. State-sponsored strategic cyber threats targeting critical infrastructure (electricity supply in particular) were prioritized by this organization [CCC 20]. The potential damage from these threats was greater than the one anticipated from cybercrime.

Apart from this information, the CCC report contains general remarks, as well as predictions, but hardly mentions any precise figures regarding cybercrime. However, the CCC informed us that ransomware largely targeted research centers and health institutions in Canada in different ways, which was corroborated by Interpol [INT 20b].

More specifically, according to the CCC, since January 2020 foreign intelligence services and cybercriminals had focused on the exploitation of the means used by individuals of strategic interest who found themselves

---

3. The CCC is the Canadian equivalent for a CERT (Computer Emergency Response Team). It acts as a first responder in cybersecurity and works in coordination with the RCMP and the Canadian Anti-Fraud Center in the fight against cybercrime.

working from home, such as virtual private networks (VPNs) or video conferencing platforms. Their goals also changed and further developed cyber espionage geared toward collecting data related to Covid-19 and its state responses in Canada, the theft of intellectual property on medical research done in the field, as well as classified information on government responses [CCC 20].

Overall, the cyber threat during the pandemic largely revolved around four main themes: ransomware targeting health and medical research establishments, online influence campaigns on Covid-19 (including many malicious impersonations of the Canadian government either online or by telephone, or by exploiting the fear generated by the pandemic), the attempted collection of data by foreign intelligence services and cybercriminals, and the various threats aimed at Canadian teleworkers. This information confirms the links between the pandemic and cybercrime.

Regarding the average ransom payments, the CCC noticed an increase of approximately 33% between the fourth quarter of 2019 (October, November and December) and the first quarter of 2020 (January, February and March), corresponding to the start of the pandemic and lockdown measures in Canada. The average payment requirement increased from 115,000 CAD to 150,000 CAD. This increase has been a notable trend since 2018 as the average at that time was 10,000 CAD. The report suggested that ransomware attacks would continue to target large enterprises and infrastructure providers and that the average ransom payment would probably continue to rise [CCC 20].

2.5. Discussion

Among the multiple dimensions of the pandemic crisis, the impact of cybercrime has been a prominent phenomenon over the past 3 years. It is interesting to evaluate whether the accelerated and sudden shift of a multitude of human activities toward cyberspace due to the pandemic crisis significantly weighed on cybercrime. Has the extent of this impact stemmed from a combined effect between cyberization, globalization, the lack of international cooperation and the unpredictability of the SARS-CoV-2 virus? Indeed, the crisis or crises caused by the pandemic greatly increased the vulnerabilities of information systems, and more specifically, that of cybersecurity and population safety in terms of overall health, throughout the world. The resilience of societies was put to the test. Canada and the United States were no exception. However, the main question of this chapter focused on how this pandemic crisis marked the evolution of cybercrime in Canada and the United States.

While there is consensus in the literature that the SARS-CoV-2 pandemic crisis increased the vulnerabilities of organizations [NAB 22] and individuals facing cybercrime [PET 21], the details of this claim still require further clarification in Canada and the United States. Obviously, there was an increase in the number of cybercrimes at the start of the pandemic crisis, contrary to what Hawdon's preliminary results heralded in Hawdon et al. in May 2020 [HAW 20]. The IC3 and IBM in the United States, like the CCC in Canada, clearly noticed it in the increase in the number of complaints recorded and an overall increase in cybercrime [INT 20, INT 21, IBM 21, CCC 20]. The various reports consulted emphasize the widening of the attack surface for cybercrimes targeting individuals due to the shift toward teleworking at the pandemic onset. Losses associated with cybercrime also increased. These increases were accompanied by an evolution of cybercrime as a consequence of the crisis. This development was noticed in the targets aimed at health research centers, hospitals and government measures (through impersonation or identity theft). Responding to the crisis became priority and potentially more lucrative targets for malicious actors. Their modus operandi also changed, increasingly resorting to ransomware and phishing. Paradoxically, even the decreasing number of cybercrimes caused more losses than it had done in the previous years. The strengthening of cybersecurity measures (two-factor authentication, for example [MER 21]) or the introduction of new measures (digital identity, for example [GRI 22]), and raising awareness about better computer hygiene could not harness those increases.

There was an undoubted increase in cybercrime and a change in the practices of malicious actors during the pandemic crisis. The targets of these criminals changed, aiming not only at the key sectors of the economies of the two states (the pharmaceutical sector, among others, and critical infrastructure) and their societies (notably, the hospital sector), but also organizations in a more destructive way than they had done in the past, through the use of sophisticated ransomware such as the Conti Group ransomware [INT 21]. Individuals, for their part, became victims of large-scale phishing campaigns set in the context of the pandemic crisis rather than of other types of cybercrimes.

Nevertheless, in this analysis, one element deserves a nuance, namely the fact that it concerns two federal states which are divided either into federated states or provinces, inducing a variation between lockdown periods, as well as in regulations and work-from-home orders and other health measures, not to mention the efforts to combat cybercrime. Thus, the reports consulted are also inaccurate, in that they provide general and unique data at a national scale for the United States and Canada, but little information on the federated levels. This results in a poor differentiation between states or provinces when disclosing cybercrime information. It is difficult to precisely pinpoint whether, in states or provinces hard hit by the pandemic, there was an increase (or not) in cybercrime compared to other regions spared by the pandemic. These considerations are conducive to general assumptions, which certainly come from the interweaving of several reliable pieces of information, but which are still imprecise to enable a fine and detailed comparative analysis.

Although law enforcement agencies in Canada and the United States cooperated with one another and with Interpol, the demographic disparities between these countries reflected in the different cybercrime statistics should not be overlooked. In other words, the United States seems to have been a more attractive target for malicious actors than Canada. As of July 31, 2020, the United States had approximately 326 million inhabitants for an Internet penetration rate of 95.6%, while the Canadian population was approximately 37 million inhabitants, with a penetration rate of 89.9% [INT 20c]. This disparity was evident in the number of cybercrime victims who filed complaints with the CCC and IC3 services. For example, in 2021, Canada recorded 5,788 victims of cybercrime, while in the United States this figure was 466,501 according to the IC3 [INT 21].

In fact, the main concern for conducting the analysis is the lack of sufficient reliable quantitative resources over the long term and on a regular basis. Indeed, the accuracy of the available data can only be computed by an annual breakdown, at best. Additionally, the categories of crimes listed may change from year to year, as was the case with the IC3 2021 report. These information gaps undermine a precise and clear analysis on the conclusions to be drawn as to the exact causes of cyber risks due to cybercrime and the cybersecurity issues observed in North America. It is difficult to discriminate, for example, whether it was the shift to teleworking or the crisis as a whole which provoked the changes observed in cybercrime, although the former is inevitably part of the latter, necessarily forcing the hazardous correlation between these two variables. Any conclusions drawn will suffer from these gaps.

This being admitted, several questions remain unanswered as to the future of the fight against cybercrime. Firstly, will the measures to combat and monitor cybercrime (de-escalation measures, prevention, etc.) during the pandemic remain as important once the crisis has been mitigated? Once the crisis has passed or is overtaken by other crises, will international monitoring and cooperation mechanisms, whether between states or between the public and private sectors, continue? Secondly, will this opportunity to cooperate between states (by means of the harmonization of policies, resource allocation, coordination meetings, etc.) leave a mark on international relations by deepening the norms and international law in the fight against cybercrime? As the crisis set a precedent, it would be logical for cooperation and collaboration to persist beyond the context of the crisis and to reduce the divergence of interests between states in the long term. One way out would be to simply get more states to ratify the 2001 Cybercrime Convention and to have it updated so as to match the incredible developments which have taken place in cyberspace [PET 21]. Thirdly, can the pandemic crisis and the reflections it generates become an opportunity to create a consensual international lexicon on the notions covered by cybercrime, including a typology of cybercrimes? Cybercrime continues to increase, year after year, both in terms of the number of victims and in terms of estimated losses. What will the tolerance threshold of governments and societies endangered by this transnational phenomenon be before the recourse to international cooperation is deemed a feasible solution? The improvement of international statistical data in view of obtaining a clear and global portrait would be a first step in this direction and a major asset in the fight against malicious actors. For the moment, in the current international context, it seems that the answers to these questions range from the improbable to the impossible [KAT 21].

2.6. Conclusion

This chapter posed the following question: how did the SARS-CoV-2 pandemic and its aftermath affect cybercrime in the United States and Canada? It is possible to conclude that, in broad terms, the pandemic crisis and the various government measures had an impact on cybercrime in Canada and the United States from the pandemic onset in North America in

March 2020 until 2022. It is also possible to affirm that cybercrime increased and was transformed, targeting sectors and institutions of interest associated with the pandemic. It took maximum advantage of the fear generated by the uncertainties of the pandemic, targeting increasingly vulnerable Internet users. However, it is difficult to draw a difference between the "natural" increase in cybercrime and the increase exclusively due to the pandemic crisis because there is a lack of concluding evidence.

To arrive at this answer, research had to face significant methodological challenges which nuance the implications of these findings. One of the methodological challenges affecting research lies in the accessibility, in the very existence of resources and the quality of data about cybercrime. In fact, it was above all the lack of reliable, regular and long-term resources which posed a problem for evaluating the phenomenon. Indeed, the reports from the Canadian and US government agencies are generally published annually. However, these reports have only little hindsight, be it quantified or qualitative, over several years, in relation to the phenomenon. This undermines the possibility of making clear quantitative or qualitative comparisons over the long term, and sometimes even, from year to year.

Not only is there a variability in the access to data, but also a variability provoked by the vocabulary used in the reports, which directly affects the quality of the data available. While cybercrime does not yet have a consensual definition at the academic level [DÉC 20], the same is true on the operational level. Although cybercrime, a cyberattack and a cybersecurity incident are all different phenomena, they are difficult to discern empirically due to a "diffuse criminological background". These different types of operations are often confused in the reports, which makes the data presented decreasingly accurate. This is, in part, not only due to the legal qualification of crimes which differs from one jurisdiction to another, from state to state, but also due to the data collection mode [DÉC 20], not to mention the potential absence of denunciations to the police services. The filing of a complaint only lists the crimes or incidents reported, whereas the data collected from sensors on the networks only focuses on the cyberattacks they are programmed to detect. Finally, there are also the states which cooperate more or less with international institutions in the fight against crime, such as Interpol. In its 2020 report on cybercrime and the pandemic, this institution – composed of 195 member states – mentioned that only 23 out of 35 states in the Americas responded to the organization's requests regarding cybercrime [INT 20b]. Another major obstacle lies in the different methods used by the responsible agencies for compiling and recording complaints. For example, IC3 reports warn that a filed complaint may include several cybercrimes, and some complaints can thus be filed and recorded several times, creating unwanted duplicates [INT 21].

In the light of this chapter, it is possible to observe an adaptation of criminal practices in cyberspace to the pandemic context, since the beginning of 2020 until present. There was also an increase in cybercrime during the same period. Government anti-cybercrime agencies and private cybersecurity firms clearly noticed this. Despite this observation, given the lack of conclusive data, it is impossible to state that the adaptation of malicious actors and their practices was solely and clearly caused by the pandemic crisis. It is just as difficult to affirm that such an increase was mainly due to the pandemic crisis. The pandemic crisis was multifactorial and in this etiological complexity, it is difficult to isolate the preponderance of the causes explaining the increase and adaptation of cyber criminality. The global and real portrait of cybercrimes and cyberattacks still remains inaccessible to research.

2.7. Acknowledgments

The author would like to thank Mr. Samuel Enright, Mr. Charles Racine and Mrs. Eva Croci for their involvement in the research for this chapter.

2.8. References

[ABS 20] ABSOLUTNET, Covid-19 et eCommerce au Québec et au Canada, available at: https://absolunet.com/wp-content/uploads/2020/03/Absolunet-Covid-19-et-eCommerce-au-Que%CC%81bec-et-au-Canada-FR.pdf, 2020.

[AHM 20] AHMAD T., "Corona virus (Covid-19) pandemic and work from home: Challenges of cybercrimes and cybersecurity", *SSRN Electronic Journal*, 2020.

[AJM 21] AJMC Staff, A timeline of Covid-19 developments in 2020, available at: https://www.ajmc.com/view/a-timeline-of-covid19-developments-in-2020, 2021.

[ALA 21] ALAVA S., "Internet est-il un espace de radicalisation ?", in Morin D., Aoun S., Al Baba Douaihy S. (eds), *Le nouvel âge des extrêmes ? : Les démocraties occidentales, la radicalisation et l'extrémisme violent*, Les Presses de l'Université de Montréal, Montreal, 2021.

[ALZ 20] ALZAHRANI A., "Coronavirus social engineering attacks: Issues and recommendations", *International Journal of Advanced Computer Science and Applications*, vol. 11, no. 5, p. 154, 2020.

[ARU 15] INHA A., NGUYEN T.H., KAR D. et al., "From physical security to cybersecurity", *Journal of Cybersecurity*, vol. 1, no. 1, pp. 19–35, 2015.

[AUT 20] CANADIAN INTERNET REGISTRATION AUTHORITY (CIRA), Canada's Internet Factbook, available at: https://www.cira.ca/resources/factbook/canadas-internet-factbook-2020, 2020.

[BAD 20] BADAWI E., JOURDAN G.-V., "Cryptocurrencies emerging threats and defensive mechanisms: A systematic literature review", *IEEE Access*, vol. 8, pp. 200021–200037, available at: https://ieeexplore-ieee-org.ezproxy.usherbrooke.ca/stamp/stamp.jsp?tp=&arnumber=9243940, 2020.

[BAN 20] BANDARA I., "Cybersecurity concerns in E-learning education", *Proceedings of ICERI2014 Conference 17th-19th November 2014*, Buckinghamshire New University, available at: http://ecesm.net/sites/default/files/ICERI_2014.pdf, 2014.

[BAU 14] BAUMARD P., "La cybercriminalité comportementale : historique et régulation", *Revue française de criminologie et de droit pénal*, no. 3, pp. 39–75, 2014.

[BEC 01] BECK U., *La société du risque. Sur la voie d'une autre modernité*, Aubier, Paris, 2001[4].

[BEN 22] BENCHERIF A. et al., Étude internationale sur les dispositifs de prévention de la radicalisation et de l'extrémisme violents dans l'espace francophone, Chaire UNESCO en prévention de la radicalisation et de l'extrémisme violents, 2022.

[BRE 10] BRENNER S.W., *Cybercrime: Criminal Threats from Cyberspace*, Praeger, Westport, CT, 2010.

[BUI 20] BUIL-GIL D., MIRÓ-LLINARES F., MONEVA A. et al., "Cybercrime and shifts in opportunities during Covid-19: A preliminary analysis in the UK", *European Societies*, vol. 2, pp. 1–13, 2020.

[CAM 22] CAMACHO S., BARRIOS A., "Teleworking and technostress: Early consequences of a Covid-19 lockdown", *Cognition, Technology & Work*, vol. 24, pp. 441–457, 2022.

[CAN 21] CANADIAN PRESS, From first cases to first vaccines: A timeline of Covid-19 in Canada, available at: https://globalnews.ca/news/7597228/covid-canada-timeline/, January 25, 2021.

---

4. For the English edition: Beck U., *Risk Society: Towards a New Modernity*, Sage, London, 1992.

[CCC 20] CENTRE CANADIEN POUR LA CYBERSÉCURITÉ, Évaluation des cybermenaces nationales 2020, available at: https://cyber.gc.ca/en/guidance/national-cyber-threat-assessment-2020, 2020.

[CEN 22] CENTER FOR INTERNET SECURITY, Cyber attacks: In the Healthcare sector, available at: https://www.cisecurity.org/blog/cyber-attacks-in-the-healthcare-sector/, 2022.

[CHA 18] CHAUDHARY T., JORDAN J., SALOMONE M. et al., "Patchwork of confusion: The cybersecurity coordination problem", *Journal of Cybersecurity*, vol. 4, no. 1, tyy005, 2018.

[COH 79] COHEN L., FELSON M., "Social change and crime rate trends: A routine activity approach", *American Sociological Review*, vol. 44, pp. 588–608, 1979.

[COL 20] COLLIER B., HORGAN S., JONES R. et al., "The implications of the Covid-19 pandemic for cybercrime policing in Scotland: A rapid review of the evidence and future considerations", *The Scottish Institute for Policing Research*, available at: https://www.researchgate.net/publication/341742472_Issue_No_1_The_implications_of_the_Covid-19_pandemic_for_cybercrime_policing_in_Scotland_A_rapid_review_of_the_evidence_and_future_considerations, 2020.

[DÉC 20] DÉCARY-HÉTU D., "La cybercriminalité", *CrimRxiv*, DOI: 10.21428/cb6ab371.9e0bdd09, 2020.

[DUP 16] DUPONT B., "La gouvernance polycentrique du cybercrime : les réseaux fragmentés de la coopération internationale", *Cultures & Conflits*, 102, available at: http://conflits.revues.org/19292, 2016.

[DUP 17] DUPÉRÉ S., "Les différentes couches composant le cyberespace" in LOISEAU H. and WALSDISPEUHL E. (eds), *Cyberespace et science politique, de la méthode au terrain, du virtuel au réel*, Presses de l'Université de Québec, Quebec, 2017.

[EUR 20] EUROPOL, Catching the Virus – Cybercrime, disinformation and the Covid-19 pandemic, available at: https://www.europol.europa.eu/publications-documents/catching-virus-cybercrime-disinformation-and-covid-19-pandemic, 2020.

[FBI 20] FEDERAL BUREAU OF INVESTIGATION, Cyber actors take advantage of Covid-19 pandemic to exploit increased use of virtual environments, available at: https://www.ic3.gov/Media/Y2020/PSA200401, 2020.

[GAY 21] GAYRAUD J.-F., "Les grandes criminalités, entre réalité géopolitique et menace stratégique", *Revue Défense Nationale*, no. 842, pp. 28–33, 2021.

[GRC 14] GENDARMERIE ROYALE DU CANADA, Cybercriminalité : survol des incidents et des enjeux au Canada, available at: https://www.rcmp-grc.gc.ca/fr/cybercriminalite-survol-des-incidents-et-des-enjeux-au-canada#sec2, 2014.

[GRC 15] GENDARMERIE ROYALE DU CANADA, Stratégie de lutte contre la cybercriminalité de la gendarmerie [sic] royale du Canada, available at: https://www.rcmp-grc.gc.ca/fr/strategie-lutte-cybercriminalite-gendarmerie-royale-du-canada, 2015.

[GRI 22] GROUPE DE RECHERCHE INTERDISCIPLINAIRE EN CYBERSÉCURITÉ (GRIC), Guide d'encadrement sécuritaire de l'identité numérique, Université de Sherbrooke, Quebec, March 2022.

[HAW 20] HAWDON J., PARTI K., DEARDEN T.E., "Cybercrime in America amid Covid-19: The initial results from a natural experiment", *American Journal of Criminal Justice*, no. 45, pp. 546–562, 2020.

[HOS 22] HOSSEINZADEH P., ZAREIPOUR M., BALJANI E. et al., "Social consequences of the Covid-19 pandemic: A systematic review", *Investigación y Educación en Enfermería*, vol. 40, no. 1, available at: https://doi.org/10.17533/udea.iee.v40n1e10, 2022.

[IBM 21] IBM SECURITY, X-Force Threat Intelligence Index 2021, available at: https://www.cert.hu/sites/default/files/xforce_threat_intelligence_index_2021_90037390usen.pdf, 2021.

[INT 20] INTERNET CRIME COMPLAINT CENTER, Internet Crime Report 2020, Federal Bureau of Investigation, available at: https://www.ic3.gov/Media/PDF/AnnualReport/2020_IC3Report.pdf, 2020.

[INT 20b] INTERPOL, Cybercrime: Impact of Covid-19, available at: https://www.interpol.int/News-and-Events/News/2020/INTERPOL-report-shows-alarming-rate-of-cyberattacks-during-Covid-19, 2020.

[INT 20c] INTERNET WORLD STATS, Internet usage, Facebook subscribers and population statistics for all the Americas world region countries, available at: https://www.internetworldstats.com/stats2.htm, 2020.

[INT 21] INTERNET CRIME COMPLAINT CENTER, Internet Crime Report 2021, Federal Bureau of Investigation, available at: https://www.ic3.gov/Media/PDF/AnnualReport/2021_IC3Report.pdf, 2021.

[KAI 21] KAI M.F., La preuve numérique à l'épreuve de la cybercriminalité, Mémoire de 2 Droit pénal international et européen, Université de Limoges, 2021.

[KAT 21] KATAGIRI N., "Why international law and norms do little in preventing non-state cyber attacks", *Journal of Cybersecurity*, vol. 7, no. 1, tyab009, 2021.

[KAT 22] KATAGIRI N., "Explaining cyberspace dynamics in the Covid Era", *Global Studies Quarterly*, vol. 2, ksac022, 2022.

[KOT 17] KOTT A., SWAMI A., WEST B.J., "The fog of war in cyberspace", *Computing Edge Security*, vol. 49, pp. 84–87, 2017.

[LAL 20] LALLIE H.S., SHEPHERD L.A., NURSE J.R.C. et al., "Cybersecurity in the age of Covid-19: A timeline and analysis of cyber-crime and cyber-attacks during the pandemic", *arXivLabs*, available at: https://arxiv.org/abs/2006.11929, 2020.

[LAW 22] LAWSON T., NATHANS L., GOLDENBERG, A. et al., Covid-19: Emergency measure tracker, available at: https://www.mccarthy.ca/en/node/63791, 2022.

[LEC 10] LECOMTE R., "L'anonymat comme 'art de résistance' : le cas du cyberespace tunisien", *Terminal, Technologies de l'information, culture & société*, no. 105, available at: http://journals.openedition.org/terminal/1862, 2010.

[MER 21] MERVENT P., "Covid-19 et télétravail : échantillon de solutions informatiques sécurisées. 1) État des lieux", *Les notes du CREOGN*, no. 59, available at: https://hal.archives-ouvertes.fr/hal-03223463, 2021.

[MIC 20] MICROSOFT 263 DEFENDER THREAT INTELLIGENCE TEAM, Exploiting a crisis: How cybercriminals behaved during the outbreak, available at: https://www.microsoft.com/security/blog/2020/06/16/exploiting-a-crisis-how-cybercriminals-behaved-during-the-outbreak/, 2020.

[MOH 20] MOHLER G., BERTOZZI A.L., CARTER J. et al., "Impact of social distancing during Covid-19 pandemic on crime in Los Angeles and Indianapolis", *Journal of Criminal Justice*, vol. 68, 101692, 2020.

[MOR 20] MORASSE M.È., Éducation : un "fossé" entre le public et le privé, available at: https://www.lapresse.ca/covid-19/2020-04-06/education-un-fosse-entre-le-public-et-le-prive, April 6, 2020.

[NAB 22] NABE, C., Impact de la Covid-19 sur la cybersécurité, available at: https://www2.deloitte.com/ch/fr/pages/risk/articles/impact-covid-cybersecurity.html, 2022.

[NOR 20] NORTHCOTT, P., La lutte contre la cybercriminalité en temps de pandémie, available at: https://www.rcmp-grc.gc.ca/fr/gazette/lutte-cybercriminalite-temps-pandemie, 2020.

[OLO 20] OLOFINBIYI S., SHANTA B.S., "The role and place of Covid-19: An opportunistic avenue for exponential world's upsurge in cybercrime", *International Journal of Criminology and Sociology*, vol. 9, 221–230, 2020.

[PAR 20] PARK A., MONTECCHI M., FENG C.M. et al., "Understanding 'fake news': A bibliographic perspective", *Defence Strategic Communications*, vol. 8, pp. 141–172, 2020.

[PET 21] PETERS A., "Is Covid-19 changing the cybercrime landscape?", in HAKMEH J., TAYLOR E., PETERS A. et al., *The Covid-19 Pandemic and Trends in Technology, Transformations in Governance and Society*, Chatham House, London, available at: https://www.chathamhouse.org/2021/02/covid-19-pandemic-and-trends-technology/03-covid-19-changing-cybercrime-landscape, 2021.

[REG 22] REGALADO J., TIMMER A., JAWAID A., "Crime and deviance during the Covid-19 pandemic", *Sociology Compass*, vol. 16, e12974, DOI: 10.1111/soc4.12974, 2022.

[SAL 22] SALEOUS H., ISMAIL M., ALDAAJEH. et al., "Covid-19 pandemic and the cyberthreat landscape: Research challenges and opportunities", *Digital Communications and Networks*, DOI: 10.1016/j.dcan.2022.06.005, 2022.

[SAN 22] SANS INSTITUTE, Glossary of security terms, available at: https://www.sans.org/security-resources/glossary-of-terms/, 2022.

[SEN 20] SENTRY BAY, Security expert predicts at least 30-40% increase in cyber-attacks during Coronavirus, available at: https://www.sentrybay.com/news/article/security-expert-predicts-at-least-30-40-increase-in-cyber-attacks-during-coronavirus, 2020.

[TAD 20] TADESSE S., MULUYE W., "The impact of Covid-19 pandemic on education system in developing countries: A review", *Open Journal of Social Sciences*, vol. 8, pp. 159–170, DOI: 10.4236/jss.2020.810011, 2020.

[TAH 21] TAHSIN M., MORISSETTE R., "Working from home after the Covid-19 pandemic: An estimate of worker preferences", *Statistics Canada*, available at: https://www150.statcan.gc.ca/n1/pub/36-28-0001/2021005/article/00001-eng.htm, 2021.

[UNE 20] UNESCO, UNESCO rallies international organizations, civil society and private sector partners in a broad Coalition to ensure #LearningNeverStops, available at: https://en.unesco.org/news/unesco-rallies-international-organizations-civil-society-and-private-sector-partners-broad, 2020.

[UNO 20] UNODC (UNITED NATIONS OFFICE ON DRUGS AND CRIME), The impact of Covid-19 on organized crime, Research Brief, available at: https://www.unodc.org/unodc/en/frontpage/2020/July/organized-crime-groups-are-infiltrating-the-legal-economy-following-covid-19-crisis--says-latest-unodc-research-brief.html, 2020.

[USB 20] U.S. BUREAU OF LABOR STATISTICS, Supplemental data measuring the effects of the coronavirus (Covid-19) pandemic on the labor market: Table 1. Employed persons who teleworked or worked at home for pay at any time in the last 4 weeks because of the coronavirus pandemic by selected characteristics, available at: https://www.bls.gov/cps/effects-of-the-coronavirus-covid-19-pandemic.htm, 2020.

[VEN 20] VENTRELLA E., "Privacy in emergency circumstances: Data protection and the Covid-19 pandemic", *ERA Forum*, vol. 21, pp. 379–393, available at: https://link.springer.com/article/10.1007%2Fs12027-020-00629-3, 2020.

[WAL 07] WALL D., *Cybercrime: The Transformation of Crime in the Information Age*, Polity Press, Cambridge, 2007.

[WAL 17] WALDISPUEHL E., BRANTHONNE A., "L'ethnographie virtuelle, quand le terrain montre les enjeux éthiques de la méthode" in LOISEAU H., WALSDISPEUHL E., *Cyberespace et science politique, de la méthode au terrain, du virtuel au réel*, Presses de l'Université de Québec, Quebec, 2017.

[WHO 22a] WORLD HEALTH ORGANIZATION, Canada, Confirmed Cases, available at: https://covid19.who.int/region/amro/country/ca, 2022.

[WHO 22b] WORLD HEALTH ORGANIZATION, United States of America, available at: https://covid19.who.int/region/amro/country/us, 2022.

[WIK 22a] WIKIPEDIA, US state and local government responses to the Covid-19 pandemic, available at: https://en.wikipedia.org/wiki/US_state_and_local_government_responses_to_the_Covid-19_pandemic, 2022.

[WIK 22b] WIKIPEDIA, Covid-19 pandemic in Canada, available at: https://en.wikipedia.org/wiki/Covid-19_pandemic_in_Canada, 2022.

3

Online Radicalization as Cybercrime: American Militancy During Covid-19

3.1. Introduction

On January 6, 2021, a swarming amalgamation of far-right militants, self-styled "digital soldiers" [FED 21] and followers of an obscure online persona known as Q, participated in a mass insurrection that targeted the United States Capitol Complex. During the attack, which has no precedent in US history, the insurrectionists sought to disturb the nation's Constitutional order, in the belief that the incoming administration of President-Elect Joe Biden was illegitimate. In the process, they vandalized the federal bicameral legislature of the United States, while parading in its corridors dressed in paramilitary apparel and waving Confederate and far-right banners. Many of them actively conspired to murder legislators and law enforcement personnel [FEU 22]. Nearly half of the over 2,000 rioters were apprehended by authorities following the most complex federal investigation in US history.

Most of the individual rioters faced misdemeanor charges for trespassing, disorderly conduct and general disruptive activity. As a political act, however, the insurrection rose to the level of a crime against the state [BES 21]. Chapter 115 of Title 18 of the *United States Code*, entitled "Treason, Sedition and Subversive Activities", defines crimes against the state as threats to the established political order by those who intend to cause the overthrow, or destruction, of the government of the United States [HEA 11].

Chapter written by Joseph FITSANAKIS and Alexa MCMICHAEL.

The attack on the Capitol marked a major culmination of a broader period of sociopolitical militancy in the United States, which has no parallel in the nation's recent history. The trajectory of that period, which in many ways has yet to subside, has closely traced the spread of the SARS-CoV-2 pandemic. Throughout this pandemic, anti-government militants have consistently portrayed themselves as opponents of state-mandated restrictions aimed at combatting the disease. These militants began acting exclusively online, some as early as late 2019. Soon, however, their activities began to metastasize in a variety of real-life domains. Thus, in April of 2020, amidst the first state-mandated lockdowns across the United States, self-styled "liberating" anti-government groups organized armed protests across the nation, with the participation of tens of thousands of militants. Some of them even plotted to kidnap public officials, sabotage rail transport systems and attack law enforcement and medical facilities with military-grade explosives [JON 20]. Others discussed plans to use Covid-19 as a weapon to target members of local and federal law enforcement with saliva or spray bottles containing bodily fluids of Covid-19 patients. Participants in a white supremacist online forum suggested smearing "saliva on door handles" at Federal Bureau of Investigation field offices, or smearing other bodily fluids on elevator buttons of apartment buildings located in "nonwhite neighborhoods" [WAL 20].

These actions came at the heels of increasingly aggressive rhetoric by domestic violent extremists (DVEs) who, in the words of the FBI, viewed the socioeconomic effects of the pandemic as a force that could potentially "crash the global economy, hasten societal collapse, and lead to a race war" [JOH 20]. Such actors are often referred to as adherents of "accelerationism", a term that dates back to the 1960s and is usually divided into distinct – though often disordered and self-contradictory – left-wing and right-wing variants. Accelerationists are proponents of the view that the capitalist system contains inherent flaws that cannot be reformed but can instead be utilized to bring down the entire system through a host of social, political and economic processes. The latter are seen as methods of accelerating the system's demise and are therefore seen as the harbingers of an age of radical social change.

3.2. A new typology of cybercrime

As will be explained in subsequent sections of this chapter, the scholarly literature is unanimous in viewing the growth of socio-political militancy in the United States as the real-world metastasis of what is essentially a cyber milieu. The unprecedented experience of the coronavirus pandemic has only strengthened this view. Researchers point to – among other variables – the striking speed with which the accelerationist rejectionism of Covid-19 mitigation measures emerged in online social-media ecosystems, such as 4chan, Reddit, Facebook and YouTube. Subsequently, in efforts to evade growing restrictions imposed by these platforms, the amorphous accelerationist movement proceeded to migrate into non-mainstream corners of the social-media universe. This led to an unprecedented period of growth for new social media platforms, such as Parler, Gab, Signal, MeWe and TRUTH Social, which practice very limited content moderation. In the ensuing period, these self-described "alternative" platforms have essentially operated as online safe havens for the recruitment, training and coordination of various typologies of violent extremists, ranging from DVEs to constitutional militia members. In short, the Internet has been operationalized by radical accelerationists as a criminogenic environment, leading to the disinhibition, mobilization and activation of large numbers of users [HUN 20]. Within this broader sociopolitical context, Covid-19 deniers are not simply finding refuge in online safe havens but are also transitioning with unprecedented speed toward real-world criminal activity. According to United States government reports, QAnon and anti-vaccine activists, who are ideologically motivated against law and order, have been actively "mobilizing against health restrictions imposed to combat the novel coronavirus" [FED 21] and are consciously transitioning from "keyboard warriors [to] engaging in real-world violence" [BER 20] and other criminal activity against the state.

This chapter posits that, although the basic analytical framework outlined above remains valid, it is insufficient to explain the magnitude of the accelerationist pressure that the United States experienced at the height of the Covid-19 pandemic. We point to the unprecedented extent, velocity and overall dynamic of accelerationist activity that confronted American political institutions during the coronavirus pandemic, and argue that they were fundamentally different to anything these institutions have experienced in recent history. Moreover, we posit that the accelerationist threat, which the United States continues to face in the post-Covid-19 environment, goes far beyond the conventional radicalization challenges posed by Web 1.0 and Web 2.0. Instead, post-Covid-19 political militancy can be more appropriately reconceptualized as a form of networked "meta-radicalism" in which persistent forms of online and offline accelerationism are becoming inseparable, spanning the digital and physical worlds in a seamless fashion. What is more, we argue that the danger of the present moment is heightened by the existence of substantial gaps in our strategic understanding of this new form of networked meta-radicalism. The latter transcends traditional criminal activity against the state, in both its cyber and offline manifestations, and poses direct conceptual challenges to our existing methodological frameworks for studying the connection between cybercrime and its real-life metastasis.

Furthermore, we propose that understanding and countering this new phenomenon requires an urgent re-examination and expansion of conventional definitions of cybercrime. The latter is traditionally understood as crime that involves the use of digital networks or computer systems. This standardized definition is well expressed by – among others – Wall, who classifies cybercrime as crime that takes place in the machine, through the machine, or against the machine [WAL 07]. We argue, however, that this conventional classification ought to be enhanced to include non-digital criminal activity – including crimes against the state – whose scope, magnitude and force are vastly enhanced through the use of the cyber domain. We posit, therefore, that the pandemic-era radicalism that America has witnessed since 2019 is the real-world metastasis of an online crime scene, which must be understood as a new typology of cybercrime and a force-multiplier for political extremism.

The idea of treating the process of online radicalization as a cybercrime is not new. It was suggested as early 2013 by scholars of Islamist radicalization, who proposed viewing online radicalization as a "prelude to real-world crime" [GRA 13] and classifying it as a cybercrime. Following in their footsteps, we argue that the crimes against the state that occurred in the United States during the pandemic were not only computer-enabled, but also computer-dependent. Thus, the unprecedented growth of political radicalism in the American context would not have been possible without the cyber domain. The Internet, in other words, was in many ways the causal element that promulgated real-life criminal activity against the state under the conditions of the pandemic. This is not to say that traditional cybercrime did not occur during the era of Covid-19. In fact, it has been claimed that fraudsters exploited the American government's Covid-19 relief plan, known as the Paycheck Protection Program, to perpetrate the "biggest fraud in a generation" [DIL 22]. Yet, the attack on the US Capitol Complex on January 6, 2021, is far broader in terms of implications and significance, and demands an extensive reconceptualization of our strategic understanding of cybercrime.

3.3. Internet connectivity and violent militancy

Radicalization is understood here as the process by which individuals evolve from the adoption of antagonistic views against people, ideologies or institutions, to actual participation in violence aimed at intimidating or coercing a civilian population [HUN 20; ALL 21; DEF 21]. It is unquestionably the case that online radicalization can be understood with reference to the methodological framework of established social-movement theory. As Bleakley [BLE 20] has shown, present-day online militancy mirrors the basic attributes seen in the militant movements of the pre-Internet era, such as the Weather Underground, a violent offshoot of the Vietnam War-era Students for a Democratic Society. He points out that militant movements of the Internet era continue to follow Jürgen Habermas' "paradigmatic shift from movements focused on economic parity towards those seeking to address non-tangible issues associated with human rights" [BLE 20]. Moreover, Internet-era militant groups continue to rely on the creation of collective environments that operate as echo chambers to form shared group identities. These identities help promote a rigid tunnel-vision approach to issues that are of concern to militant groups. As Khalil posits, online militant settings "function similarly to real-world social spaces in that they can provide identity, validation, community and meaning" [KHA 21]. Moreover, online militant groups employ these settings as propaganda platforms to attract various levels of support and to "shap[e] wider public perceptions of the movement" [BLE 20].

The above trends are hardly new. They were first observed by social science researchers as early as the mid-1990s, when militia movements in the United States were found to be utilizing interconnected computer systems to build "an imagined network community [...] and expand" [ZOO 96]. Yet, recent research has conclusively demonstrated the existence of a significant positive correlation between the increase of Internet connectivity and the growth of domestic militancy in post-industrial democracies

[HUN 20]. Broad trends in the relevant scholarly literature indicate that increased Internet connectivity in democracies tends to exacerbate pre-existing social tensions between groups in civil society, as well as amplify previously held grievances against state institutions. In the process, "political ideologies can become strengthened to the point of extremism" [HUN 20]. The broad consensus among researchers is that the Internet occupies a central function in militant recruitment and radicalization, by providing the "crucial infrastructure" [HUN 20] needed to produce a "collective chemistry" [VEL 21]. The latter helps build and bond a critical mass consisting of existing adherents of militant movements and potential new recruits. The mathematical basis of this relationship rests on the aggregation of empirical data that span an impressive volume of case studies from across the world [VEL 21]. On top of this well-established relational component of the Internet as a whole, research has identified the rising role of social media, including online gaming platforms, file-uploader applications, and end-to-end encrypted communications systems. These have been found to play "an increasingly important role in the radicalization processes of [United States] extremists *exponentially*" [YAT 16, emphasis added; see also NAT 21].

A major component in the process of radicalization relates to the mechanisms through which online echo chambers manufacture and amplify fundamental attribution errors that employ circular logic. These are primarily embodied in ideations of varying complexity which, in specialist literature, are referred to as "conspiracy theories". This often-misused term refers to the belief that human institutions are consciously designed by forces that stand to benefit from them. Conspiracy theories revolve around "ingesting and divesting cultural matter seldom of their own making" [SPA 00], in ways that help mold a uniform conspiratorial worldview. It is worth noting that, despite its relevance to today's Internet ecosystem, the term "conspiracy theory" is hardly new, having been coined by the philosopher Karl Popper in 1945 [MEL 21]. Generally speaking, "[t]here is little systematic evidence to show that the world is more conspiratorial now than it was prior to the advent of the Internet" [STA 21]. However, the research literature is in broad agreement with the view that the social media environment increases the "spreadability" of media messages promoting conspiracy theories, "with consequences on the public's belief in such theories" [USC 18].

3.4. The pre-pandemic domestic threat landscape

The onset of Covid-19 found America's domestic threat landscape in a state of heightened anxiety, following the 8 years of Barack Obama's presidency and the emergence of the phenomenon of "Trumpism" [SAV 19]. The most organized and – by far – most ominous elements of that threat landscape consisted of the heavily armed constitutional militias. American federal and state laws classify militias into organized and unorganized ones. Organized militias include members of the United States National Guard, or established military units. Unorganized militias may be composed of American citizens who are not members of organized military units but can still be called to assist in times of national emergency, or war [ADL 20]. These provisions have given rise to so-called "private" or "constitutional" militias – namely armed groups that are not recognized as organized by federal or state governments. Although the constitutional militia movement emerged in the 1990s, it traces its ideological roots to the so-called Minutemen militias of the civil-rights era, which were informed in almost equal measure by anti-communist and pro-segregationist ideology [DRA 07]. Although the constitutional militias of today do not openly subscribe to white supremacy, they can be grouped into what is often referred to in the specialist literature as the anti-government "patriot movement". Adherents of this loosely affiliated movement are united in the conspiratorial view that United States and international elites – to include multinational corporations and supranational organizations like the United Nations – are secretly building a tyrannical global government, referred to as the "New World Order" [GAL 00]. Such groups have been known to consciously direct their recruitment efforts at military service members, veterans and law enforcement personnel [WAT 21], as this tactic potentially augments their tactical capabilities. A prominent contemporary representative of the patriot movement is the Oath Keepers organization (founded in 2009), which focuses almost exclusively on recruiting current or former members of the military, law enforcement and emergency first responders communities [CSI 21]. The Oath Keepers assumed a leading role in armed marches across the United States during the coronavirus pandemic. Several of their members were subsequently charged by the government with "seditious conspiracy and obstruction of Congress" [DRU 22].

Another constitutional militia that took armed action against the state during the height of the Covid-19 pandemic is the Three Percenters. Founded in 2008, in direct response to the election of Barack Obama to the presidency of the United States, the group is a leading proponent of the belief in the inevitability – or even necessity – of a "Second American Civil War", also known as "Civil War 2.0" [BUE 21]. This is the view that the United States must, or is highly likely to, enter a period of sustained political violence or insurrection, which could progressively escalate into a large-scale armed confrontation between rival organized groups [EAT 21; GRE 21]. Several of their members were indicted for participating, military-style, in the January 6 insurrection [BUE 21]. The group joined the insurrection alongside the Proud Boys, a constitutional militia that operates based on a decentralized model throughout the United States and Canada [KUT 20]. Both the Proud Boys and the Three Percenters have been designated as terrorist organizations by Public Safety Canada [PUB 22].

Leftwing militias do exist, though they are a relatively recent phenomenon in the United States [CSI 21]. They include Redneck Revolt, the Not Fucking Around Coalition (NFAC) and a shifting constellation of John Brown Gun Clubs. Among these, the NFAC is closest in structure, tactics and strategic outlook to the right-wing constitutional militias. Its leadership consists of ethno-separatists with secessionist views, many of whom have former military experience. The group has organized several armed marches in locations like Louisville, Kentucky and Stone Mountain, Georgia [WAL 21]. Organizations like the NFAC are surrounded by largely amorphous and decentralized militant leftist and anarchist affinity groups. These include the Black Bloc, Antifa, as well as other loosely organized direct-action clusters of activists, who have militant tendencies [JON 21]. The scholarly literature is in broad agreement with the view that left-wing militant groups are substantially smaller in size, lack in organization, have inferior tactical capabilities, display a much more limited utilization of firearms and present a lower overall level of threat than right-wing constitutional militias [JON 20; COU 22].

3.5. The domestic threat landscape of the pandemic

Since 2016, right-wing constitutional militias have systematically portrayed themselves as the armed nationalist vanguard of a broader populist movement headed by the former president of the United States, Donald Trump [LYO 17]. However, it would be a strategic misconception to view the right-wing domestic threat landscape in the United States as an armed wing of the Republican movement, or as commanded in any meaningful way by former President Trump. In fact, in the weeks following the failed January 6 insurrection, senior members of the constitutional militia movement leveled harsh criticism on Trump for failing to launch a fully fledged military coup. In one characteristic instance, members of the Proud Boys decried Trump as a "weak" former president who would "go down in failure" [FRE 20]. According to the Southern Poverty Law Center, cracks in the uneasy alliance between Trump and the white supremacist wing of the patriot movement began to appear as early as the first two years of his administration [HAT 18], and only intensified in later years. There is no evidence that the coronavirus experience helped deepen the ideological convergence between the patriot movement and the Trump wing of the Republican Party. In fact, during the coronavirus period, the Internet interacted with the patriot movement in ways that made its structure more horizontal, its politics more transient and its political allegiance to established public figures – including that of Trump – far less assured.

The militant tendency that most adequately embodies the shifting contours of the American domestic threat landscape during Covid-19 is the so-called "Boogaloo movement", also known simply as "Boogaloo". The ideological identity of the Boogaloo movement is far from crystalized [KRI 21]. In its essence, the movement is reflective of a highly horizontal and decentralized ultra-libertarian point of view, which brings together a strikingly diverse range of adherents with varied political identities ranging from the far-right to the far-left. As a social movement, the Boogaloo traces its roots to the racist skinhead and neo-Nazi culture of the 1980s and 1990s, which was formed mostly inside prisons. In the early 2000s, the racist skinhead movement attempted to enter the mainstream and attract a more educated cohort – an effort that gradually birthed the so-called "alternative right", or "alt-right" movement [MAI 18]. The alt-right largely imploded after the bloody 2017 Unite the Right rally in Charlottesville, Virginia. However, for a number of years prior to 2017, some of the younger, online-savvy members of the alt-right had been active in a host of online forums revolving around gun-culture and video games [KEN 21]. It was this cross-pollination between supporters of the alt-right and online communities that birthed the Boogaloo movement. In subsequent years, that loose amalgamation of – mostly teenage and male – Internet users began to assume a recognizable ideological shape on 4chan in the weapons-themed /k/ and the racism-themed /pol/ sections of the imageboard website [KEN 21]. By 2014, these users were facing mounting countermeasures, imposed on them by the 4chan moderators due to their violent memeography. They therefore began to migrate to other social media platforms, including Twitter, Facebook, Reddit, and – especially – Discord. It was there where the anti-government themes of the Boogaloo underwent a process of gamification – the infusion of real-life concepts with videogame themes [FIN 20].

By 2015, the Boogaloo had assumed the basic elements of its identity. These include references to cult films – notably, the name of the movement is derived from the 1984 cult American musical comedy *Breakin' 2: Electric Boogaloo*. They also feature a surrealist aesthetic that pairs aloha shirts with semi-automatic rifles, as well as quirky, low-quality imagery relayed via Graphics Interchange Format (GIF) animations [CON 20]. The politics of Boogaloo have been aptly described as "apocalyptic" [EVA 20]. They center on the view that centralized government in the United States must be brought down through armed action by citizens. It follows that national or international developments that ignite chaos, bring society closer to a civil war, or otherwise hasten the collapse of the system, are to be encouraged. The ultimate desired outcome for Boogaloo adherents is a "race war [that will] facilitate the eventual collapse of the current government and replace it with a white-dominant state" [ONG 20]. This is not an absolute, however, as the Boogaloo is politically fluid and highly decentralized. Notably, there are non-racist Boogaloo groups that "rail against police shootings of African Americans, and praise black nationalist self-defense groups" [EVA 20]. Beginning in late 2019, Boogaloo adherents began to largely set aside their disparate political backgrounds and coalesce online in order to deliberate on the pandemic. Studies show that, by the spring of 2020, Boogaloo-themed social media groups were promoting a view of the coronavirus as a potentially disorienting force that could hasten society's collapse [KRI 21]. Actions taken by Boogaloo adherents against state-mandated measures to combat Covid-19 were largely crowdsourced, and resulted in some of the earliest armed presence in Covid-19-denier rallies across the United States [KRI 21].

The rise of the Boogaloo closely parallels that of another amorphous and quickly metastasizing political movement, known as QAnon. Like the Boogaloo, QAnon traces its origins in 4chan's /pol/ imageboard. It was on that imageboard where, in 2017, a user calling themselves "Q" – ostensibly with reference to the United States Department of Energy's Q clearance level – began to share alleged insights about a government conspiracy involving the former Secretary of State Hillary Clinton [AMA 20]. In subsequent posts, "Q" continued to share messages, known among QAnon adherents as "drops", purporting to reveal the existence of a secret cabal inside the government of the United States [ROT 21]. According to QAnon adherents, this cabal consists of Satan-worshiping elites who engage in the sex-trafficking of children, while also using them to extract epinephrine, a psychoactive substance that keeps them eternally young [FRI 20]. The QAnon conspiracy narrative culminates in "the Storm", also known as "the Great Awakening" or "the Event", an apocalyptic act of warfare in which members of a "QArmy", led by Donald Trump, will arrest en masse, and then physically exterminate, all members of the cabal, in an act of mass murder [FRI 20]. Following the onset of the Covid-19 pandemic, the QAnon conspiracy theory gained substantial traction among members of constitutional militias and anti-government groups, including the Boogaloo [KRI 21]. Before long, QAnon adherents began appearing in anti-mask and anti-vaccination rallies, sporting military-style gear complete with "QArmy" badges and references to the QAnon military motto, WWG1WGA ("where we go 1 we go all") [KRI 21].

3.6. Pandemic accelerationism

These online militant movements are ideologically disparate. They are, however, fueled by a common range of ideas, which broadly fall under the social theory and strategy of accelerationism. Elements of this theory are rooted in the work of the German philosopher Karl Marx, who espoused the view that the capitalist system of economic organization contains inherent contradictions. According to Marx, these contradictions are fatal and will inevitably lead to the system's eventual collapse, regardless of the intentions or efforts of those who live under it [WAL 21]. These ideas resurfaced in France following what became known as May 1968 – a 50-day-long period of widespread civil unrest, which included general strikes, riots, occupations and rallies led by leftist student groups and trade unions. The nationwide unrest resulted in the strengthening of France's centrist government, rather than – as its initiators had hoped – a revolutionary transformation of society. In its aftermath, pro-May 1968 French philosophers, Gilles Deleuze and Felix Guattari proposed that, rather than resisting a seemingly insurmountable capitalist system, its detractors should seek to hasten its maturation, and in doing so expose its fatal flaws. Thus, "[p]ushing forward is posited as the only way to push against" [LAU 17]. The hope was that society could finally rise up "only after hitting rock-bottom" [CUL 21]. The views of Deleuze and Guattari found their way into the English-language scholarly literature in the mid-1990s through the work of British philosopher Nick Land and the Cybernetic Culture Research Unit at Warwick University. Land's view was that neoliberal capitalism was so all-encompassing that it "no longer had an outside" [LAU 17] from which its detractors could resist it. This resulted in a state of "political, cultural and theoretical paralysis" [LAU 17] of radical social movements. However, capitalism could be decisively combated by radicals willing to help accelerate the inherent contradictions of the system.

Although accelerationism includes elements of traditional radicalization, it signals a drastic departure from traditional militant – or even terrorist – strategies: rather than using violence to alter a population's perspective, it seeks to use pre-existing social tensions as fissile material, in order detonate the entire social system and reduce it to mere rubble. An example of accelerationist politics is the United Kingdom's exit from the European Union, known as Brexit, which occurred in stages between 2017 and 2020. Accelerationists supported Brexit, sometimes violently, not because they believed it could improve life in Britain, but because they believed it had the potential to result in the exact opposite – and in doing so, hasten the country's collapse. In other words, accelerationists saw Brexit, not as a pathway to the future, but rather as a staunch refusal of the future – a method to coerce the system to quickly progress to its logical, self-destructive end [NOY 2].

In the work of Land, as well as Deleuze and Guattari, the principal idea behind the strategy of accelerationism centers on the participation of radical activists in real-life violent activities that can enflame existing social tensions and lead to civil disorder.

In recent years, accelerationism as a revolutionary strategy has largely migrated from left to far-right circles. Today there are growing numbers of influential far-right ideologues that promote accelerationism as a cohesive revolutionary tactic in online manifestos and blog posts [WAL 21]. These far-right ideologues agree with left-wing accelerationist interpretations of capitalism, which dismiss it as fundamentally irredeemable and unworthy of reform [MIL 20]. They therefore push for the calculated participation by far-right activists in violent actions that can potentially exacerbate social tensions and help hasten various levels of social, political and economic collapse [WAL 21]. But they differ from left-wing accelerationism in that they view capitalism's collapse as the first stage in the emergence of an authoritarian political system, infused with elements of apartheid designed to establish and maintain white supremacy [MIL 20]. Leading practitioners of American far-right accelerationism include the Base, a white nationalist paramilitary group, which has an established international presence in Russia, Canada, Australia and South Africa, among other countries [ALC 21]. A closely related far-right accelerationist group is the neo-Nazi Atomwaffen Division (AWD), also known as the National Socialist Order, which is widely considered as the most dangerous and lethal neo-Nazi organization in existence anywhere in the world [OBR 17].

3.7. From virtual to real-life criminality

Accelerationism views the development of digital communications technologies – especially networked technologies – as dialectic elements in the capitalist system. These, especially in their social media variety, are seen as lucrative distributive weapons that contain the seeds of capitalism's destruction. For accelerationists, social media sites are seen as instrumental harbingers of a radical corruption of our social structures. This is because they embody some of society's most extreme tendencies: frantic speed, overstimulation and the "data-hungry dystopia" of anarchic globalization [DIX 17]. By its very essence, the social media universe embodies unmanageable velocity, data-gigantism, social abstractness and detachment – in short, uncontrollable discord. This places them in direct opposition to what accelerationists refer to as "folk politics", namely efforts to "secede from the political field of globalized, developed capitalism. Associated with this are calls for a return to authenticity, to immediacy, to a world that is transparent, human-scaled, tangible, slow, harmonious, simple and everyday" [DIX 17]. These are dismissed by accelerationists as romanticized attempts to prolong capitalism's existence. They are therefore seen as stumbling blocks that delay the system's inevitable collapse and prevent the dawn of a post-capitalist future.

Some theorists of 21st-century accelerationism go as far as to suggest that it is built on the premise of the Internet, especially social media, which constitutes its more user-friendly and addictive component. The accelerationists of today have married traditional accelerationist ideas with the proliferating dynamic of the Internet, and especially its social media components. They stipulate that accelerationism can be facilitated through small-scale individual actions that can be propagated en masse through social media platforms. The latter are seen as having the potential to augment polarization along economic, racial and social lines in unprecedented ways, and eventually help bring about societal collapse. Consequently, the disruptive dynamics of social media are being systematically instrumentalized by accelerationist groups. During the Covid-19 pandemic, accelerationists used social media as a "means of narrative dispersion" [KRI 21] of conspiracy theories. The extensive reach of the social media universe facilitated the dispersion of conspiracy theories among previously disparate groups of actors. Researchers have termed the resulting concoction of anti-government, conspiracy-prone and gun-enthused users, "militia-sphere" [FIN 20]. The militia-sphere connects more established accelerationist elements with mainstream segments of the user population [ISD 21]. The hybridity of this emerging threat landscape provides accelerationist groups with unprecedented access to expansive audiences [ISD 21]. Moreover, accelerationist actors understand the tactical usefulness of social media as instruments that can mobilize large numbers of adherents for high-impact activities in the offline domain [FIN 20; ONG 20]. The specialist literature clearly shows that online militant groups consciously seek to demonstrate that they represent "more than just a collection of 'keyboard commandos' operating only in the virtual space" [ARY 20]. That tendency was clearly visible during Covid-19. There is also substantial consensus among researchers that social media platforms "support, encourage or mobilize real-world harm" [KHA 21] and that "a strong online dimension to offline mobilization constitutes a unifying trend" across all militant movements, ranging from the far-left to the far-right [ISD 21; WAL 21]. With reference to the United States, the utilization of social media platforms as amplifiers of accelerationist rhetoric is largely responsible for an unprecedented rise in real-world terrorist violence – primarily from the far-right – in recent years [KHA 21].

3.8. Online radicalization during Covid-19

It is important to stress here that some of the trends outlined above were identified in the literature long before the onset of the coronavirus pandemic. Indeed, militants have long used centralized, heavily vetted and encrypted online platforms for command and control functions, while also distributing propaganda on easily accessible social media platforms in order to attract new followers. That is a time-tested model for online militancy that predates our century [MIL 20]. Bleakley, in particular, has explained the process by which online forums, such as Reddit, can be used to build virtual ecosystems that separate users "from mainstream society on a philosophical level, if not a physical one" [BLE 20]. This process includes the creation of a shared set of language conventions and terminology, which help individual users construct a sense of fraternity with each other and establish a group-based "normative philosophy" [BLE 20]. Such "virtual safe havens" [NAT 21] are absolutely central to the concept of radicalization. Upon initial inspection, the process of radicalization that led to the storming of the United States Capitol in January of 2021 followed the same steps that researchers have identified in the past, in relation to DVE radicalization [NSC 21].

We posit, however, that the unprecedented social, political and economic conditions of the pandemic, coupled with the use of social media as virtual safe havens for militants, altered the nature of online radicalization in at least four fundamental ways:

– first, the social isolation imposed by the pandemic made Internet users susceptible to deep deprogramming by militant actors;

– second, there was a mass migration of users toward a growing milieu of social media platforms that reject the very principle of content moderation, with potentially far-reaching consequences;

– third, the range of the militia-sphere across all platforms grew to unprecedented proportions, allowing for a striking array of heterogeneous extremist groups to converge and build ideological confluence;

– finally, the above factors contributed to a new paradigm of online radicalization, whose accelerated speed and size appears to be facilitating the radicalization of the American population on a mass scale.

It is not hyperbole to suggest that, beginning in the spring of 2020, the coronavirus pandemic radically disrupted traditional modes of socialization in post-industrial societies. Throughout that period, as millions of Americans spent increasing amounts of time at home, they relied on the Internet as never before. Studies show that, during the first year of the pandemic, the daily in-home data usage increased by nearly 40 percent on average among American households [MCC 21]. Additionally, nearly 90 percent of Americans indicated in surveys that the Internet was at least "important" to them during the pandemic, with close to 60 percent indicating that it was "essential" [CHR 20]. Millions turned to the Internet in attempts to pierce through the long bouts of isolation, or in an effort to find financial assistance following long periods of unemployment. Among others, Walther and McCoy have explained that, as millions sat in isolation facing the screens of cellphones and computers, it became "a great deal easier for online recruiters to attract more people and expose them to disinformation campaigns" [WAL 21]. But the problem is potentially far deeper: traditional radicalization theory places much emphasis on social self-isolation, seeing it as instrumental in enabling the "deprogramming" of individuals. This is essentially seen as a form of conditioning that pushes individuals to carry out acts of violence [BLE 20]. During the pandemic, with at-home audiences held hostage by the effects of the pandemic in their millions, it was no accident that "sensational media narratives and conspiracy" theories were found to be "growing, and in some cases more than doubling across web communities" [FIN 20].

In fact, it appears that, far from being passive victims of recruitment, a considerable segment of Internet users actively sought out conspiratorial narratives of the pandemic, and were prepared to migrate to online platforms that allowed them to "not simply reject science", but to "invent [their] own scientific rhetoric" about public health [CHR 20]. This led to the growth of the so-called "alt-tech" universe [FRE 20], a constellation of self-described alternative social media platforms like Rumble, Telegram and Parler, which built their entire business model on their minimalist approach to content moderation [NAT 21]. The promotion of these alt-tech platforms by influential figures on the conservative and far-right political spectrum [WAL 21] led to what researchers have described as "the great scattering of extremists and extremist groups across alternative platforms" [JAR 22]. The latter thus rapidly became the technological support infrastructure of the far-right [DON 19] and facilitated the exponential growth of radicalized communities, including QAnon and the Boogaloo [MCS 21]. In their insightful study of the alt-tech universe, Donovan, Lewis and Friedberg metaphorically refer to it as a series of parallel ports, which allow "for multiple streams of data to flow simultaneously" [DON 19]. By facilitating that role, the alt-tech universe illustrates how "networked social movements [...] are both structured by and structuring their own technological infrastructure", the authors note [DON 19]. Certainly not all militants migrated to the alt-tech universe. Yet, even those who stayed behind found ways to evade regulation protocols on mainstream social media platforms, primarily by self-modulating their messages in creative ways. In one of many characteristic cases, Boogaloo groups began referring to themselves as

"Big Igloo" in order to avoid being targeted by automated moderation systems [KRI 21].

The "great scattering" of extremists had a twofold effect: first, it enabled extremist online content to be distributed away from the mainstream; second, it brought previously niche extremist communities in contact with elements of the larger anti-government movement. It did so largely through promoting "shared attitudes toward [doubting] the legitimacy of the pandemic, lockdown orders and the role of law enforcement and other government officials" [FIN 20]. By migrating to the alt-tech ecosystem, accelerationists interacted with unprecedented intimacy with online communities of firearms enthusiasts, survivalists, QAnon adherents and others, comparing grievances and building a "big tent" [KIL 21] for a mass movement. Historically, an entanglement of such a massive scale, and involving such varying political backgrounds and ideologies, would have prevented what extremism researchers refer to as "ideological coherency" [ISD 21]. Recent research, however, shows that the ideological antagonism that has traditionally plagued the American far-right is eroding to some extent [STA 11]. This ideological convergence is not surprising, given the invigorating effect that the coronavirus pandemic has had on online extremist communities – especially those on the anti-government spectrum [KRI 21]. Research conducted during the pandemic showed that far-right militancy proliferated with unprecedented speed on social media, such as Telegram, with Boogaloo-related chatter nearly doubling during quarantine periods [ONG 20; DAV 21]. This may be behind the notable quickening in radicalization timelines – that is, the time it takes from initial exposure to extremist beliefs to participation in extremist violence [YAT 16] – which researchers warn has changed markedly since the advent of the Covid-19 pandemic. Research from as early as 2016 has shown that increasing interaction on social media tends to speed up the duration of radicalization. Over time, the duration has shrunk from around 18 months in 2005, to as few as 7 months today [STA 11].

One is also confronted by the issue of numbers. Arguably, the most notable difference between the Covid-19 era and prior eras of domestic violent extremism in the United States relates to the sheer size of radicalized populations. On January 20, 2021, Americans were given a first taste of the size of armed radicalism during Covid-19. On that day, the Virginia Citizens Defense League, a state-based gun rights advocacy group, organized Lobby Day, a pro-gun rally, at the Virginia State Capitol in Richmond, Virginia. Over 20,000 armed protestors participated in the rally, including members of the Boogaloo [KRI 21]. Prior to the rally, and in light of reports that constitutional militias and other anti-government groups intended to cause violence, Virginia Governor Ralph Northam declared a state of emergency. A few days before the event, the FBI arrested three heavily armed members of the aforementioned neo-Nazi group the Base, who had allegedly planned to engage in violent acts at the rally [WIL 20]. That event was dwarfed by the January 6, 2021, attack on the United States Capitol Complex by tens of thousands of people, at least 2,000 of whom broke into the main building and attempted to subvert the United States Constitutional order [MCS 21]. In what is an apt illustration of the inability of social media platforms to monitor and regulate user content, plans for the attack had been openly discussed online by extremists, who had "for weeks repeatedly expressed their intentions to attend the January 6 protests, and unabashedly voiced their desire for chaos and violence online", according to news reports [LYT 21]. Despite its shocking levels of violence, the attack was but the culmination of a broader radicalized movement that questions the legitimacy of President Trump's successor, Joe Biden. Australian-American counter-insurgency expert David Kilcullen illustrates the sheer mathematical size of this problem:

> Roughly 74 million people voted for Donald Trump; a mid-November Rasmussen poll suggested that 75 percent of them believe Democrats stole the election. That's 55.5 million people who believe a Biden administration is illegitimate; if only 3 percent of them decided to act on that belief, that would be 1.66 million people, a sizeable base for revolt [KIL 21].

Kilcullen further notes that "stolen elections – or the perception of them [...], which amounts to the same thing – are some of the most common triggers for revolution" in political science research [KIL 21]. Experts encapsulate this with the use of the term "mass radicalization", a phenomenon in which large swathes of a population become directly vulnerable to extremist messages in a relatively short period of time [STA 11].

3.9. A new methodological paradigm for online radicalization?

It is difficult to dismiss the view that the onset of Covid-19 provided militant groups with a fresh narrative – a new rallying point that generated both momentum and growth in online communities [VEL 21]. However, the ensuing "cognitive war" [ROC 17] that we can observe in the rise of the Boogaloo, QAnon and other pandemic-era militant entities rests on far more than momentum. Indeed, the convergence of extremist actors [ISD 21], which defies attempts to classify them along traditional political lines, is potentially a direct product of the Internet itself, which practically encourages the production of conspiracy theories on a mass scale. In the words of Alasdair Spark, the Internet has allowed

> the potential connection of any item, any datum, to any other producing latent connections which produce transitory, temporary, meanings out of the mass. As such, it models the contemporary use made of conspiracy theories by many [SPA 00].

Moreover, the unfolding evolution of the Internet, as typified by social media culture, is moving toward less – not more – centralization. Researchers of radicalism have noted that this emerging digital landscape adds potentially insurmountable barriers to the regulation of online extremism. In particular, the dawn of encrypted peer-to-peer social media platforms, such as Telegram, makes it near-impossible to appeal to any type of centralized authority in order to contain illicit or radical communications content [MIL 20]. This makes the online ecosystem far safer for purposes of command-and-control. Moreover, it applies not only in comparison to previous stages in the Internet's evolution, but also in comparison to the offline environment, in which militants risk the potential of being doxed by rival activists [MAR 20]. Even on those social media platforms where strict content moderation is imposed, the highly adaptive and allegoric nature of pandemic-era radical movements allows them to routinely evade detection. The example of the meme-based hybrid culture of the Boogaloo is paradigmatic here: in this case, content moderation efforts have failed to notice the dual-use meaning of "harmless" memeography such as that of Pepe the Frog, green unicorns or Campbell Soup Company cans [ELL 20; KRI 21].

Inevitably, the online-infused fluid hybridity of pandemic-era radical movements is influencing their structure in fundamental ways. Not only are they far less hierarchical than pre-pandemic extremist groups, but oftentimes they are also devoid of ideological coherence, which makes it impossible to categorize them [ISD 21]. The absence of leaders, or even prominent figures, is a near-universal characteristic of accelerationist groups. The concept of leaderless resistance is manifestly not new, having first emerged out of

American far-right circles in the 1990s [BEL 18]. There is ample evidence that accelerationist strategists are systematically encouraging operational autonomy among militants, especially after the January 6 insurrection, which prompted the full weight of law enforcement to be placed on DVEs [JAR 22]. Yet, the principle of decentralization appears to be so deeply integrated into pandemic-era accelerationist doctrine that it goes beyond leaderless resistance. The latter assumes the existence of ideologically congruent cells of militants acting independently in a localized milieu [BEL 18]. However, what we have seen in post-pandemic accelerationism negates the very existence of a cell structure. Instead, violent attacks are carried out by individuals who have "no, or very loose, connection to specific organizations [and] draw instead on a shared culture and ideology" with an online forum or group [DAV 21]. Thus, for these "logistically autonomous individuals" [DAV 21], "a confluence of ideological affinities is more powerful in inspiring and provoking violence than the hierarchical terrorist organizational structures of the past" [HOF 20]. Some experts are referring to this new paradigm as "post-organizational violent extremist mobilization" [DAV 21], a term that rejects the traditional distinction between a militant's online consumption of extremist information and physical connections to real-world groups [GUH 21]. Radicalization, therefore, is not seen as a result of specific political messages found online, but as the aggregated outcome of "the giant connected component of the network [...], the gel forming from the background pool of users on the Internet" [VEL 21].

3.10. Conclusion: meta-radicalization as cybercrime

Even before the onset of Covid-19, radical accelerationists had instrumented the Internet as a criminogenic environment, in efforts to disinhibit, mobilize and activate online users. Their goal has always been – and remains today – to facilitate the transition of online militancy into real-world criminal activity. The purpose of this chapter has been to question the extent to which the basic analytical framework of cybercrime theory remains valid under pandemic conditions. Our argument is that the unprecedented accelerationist pressure that the United States experienced during Covid-19 forces us to seek new ways to understand online radicalization as a form of cybercrime. Indeed, the extraordinary extent, velocity and overall dynamic of accelerationist activity that has confronted age-old American institutions in recent years points to a new type of symbiotic association between cyber and offline elements.

Clearly, America's domestic threat landscape predates the pandemic. However, the interaction of Covid-19 with the cyber domain gave unprecedented vigor to a new breed of hybrid militant movements, such as the Boogaloo and QAnon. Although ideologically disparate, these movements are jointly fueled by a common belief in accelerationism. A theory as much as a strategy, accelerationism signals a drastic departure from traditional militant – or even terrorist – strategies in a number of important ways. Notably, it views the development of digital communications technologies – especially social media – as dialectic elements in the capitalist system, which can be utilized in highly disruptive ways. In comparison to their pre-pandemic progenitors, the resulting militant movements, such as the Boogaloo and QAnon, are more politically eclectic, more difficult to monitor, and far larger. This dynamic is largely an outcome of these movements' post-organizational logic, which is cyber-infused and is influencing their outlook and structure in fundamental ways.

What does this mean for our contemporary understanding of cybercrime? In our view, the political radicalism of the pandemic era is inconceivable outside of the cyber domain. Not only does the radicalism we witnessed under Covid-19 point to the influence of the social media dynamic, it also challenges the traditional dichotomy between cybercrime and offline activity, on which traditional definitions of cybercrime have relied in the Internet era. Before the pandemic, radicalization researchers stressed repeatedly that "[t]he Internet has not led to a rise in terrorism. It is largely a facilitative tool; radicalization is enabled by the Internet rather than being dependent upon it" [GIL 15]. The experience of the pandemic directly challenges that view. In her pandemic-era overview essay on the role of technology in violent extremism, Lydia Khalil echoes Rita Katz in asking a question, which we view as central to this analysis: are we perhaps witnessing an emergence of a new typology of militancy, one that cannot possibly exist without the cyber domain? [KHA 21]. Our answer is that we are, and that this phenomenon has emerged as a result of a new typology of cybercrime. In fact, the Capitol attack of January 6, 2021, may have been an early representative of this new typology of cybercrime.

Our argument, as outlined in this chapter, is that cybercrime during the pandemic era is, at the very minimum, substantively different from that of the pre-pandemic era. Far from simply facilitating radicalization, the social media environment has "fundamentally changed the conditions around social interaction and reorganized our public sphere in such a way that has led to extremism" [KHA 21]. This extremist logic forms the essence of 21st-century accelerationism. What we are now witnessing, therefore, is a form of what can be described as meta-radicalization – a cybercriminal typology of radicalization that blurs the traditional heterogeneity of online and offline activity. This new form of radicalization, a product of the digitization of life under Covid-19, prompts us to understand cybercrime and offline militancy as an interconnected, experiential process that seamlessly spans the digital and physical worlds around us. Undoubtedly, the forces we describe in this chapter are still taking shape. They do, however, offer us initial glimpses of what the future may hold.

3.11. References

[ADL 20] ADL, The Militia Movement, *Anti-Defamation League*, available at: https://www.adl.org/resources/backgrounders/the-militia-movement-2020, 2020.

[ALC 21] ALCORN W., *Moving Beyond Islamist Extremism: Assessing Counter Narrative Response to the Global Far Right*, Ibidem-Verlag, Stuttgart, 2021.

[ALL 21] ALLINGTON D., *Conspiracy Theories, Radicalization and Digital Media*, International Centre for the Study of Radicalisation, King's College, London, 2021.

[AMA 20] AMARASINGAM A., ARGENTINO M.A., "The QAnon conspiracy theory: A security threat in the making?", *CTC Sentinel*, vol. 7, pp. 37–44, 2020.

[ARY 20] ARYAEINEJAD K., BERGER J.M., ARYAEINEJAD K., LOONEY S., "There and back again: How white nationalist ephemera travels between online and offline spaces", *The RUSI Journal*, vol. 165, no. 1, pp. 114–129, 2020.

[BEL 18] BELEW K., *Bring the War Home: The White Power Movement and Paramilitary America*, Harvard University Press, Cambridge, MA, 2018.

[BER 20] BERTRAND N., DHS warns of increase in violent extremism amid coronavirus lockdowns, *Politico*, available at: https://www.politico.com/news/2020/04/23/dhs-increase-in-coronavirus-inspired-violence-205221, 2020.

[BES 21] BESHAW C., FRANK M., "Overview: Was the Capitol riot a coup attempt?", in WIENER G. (ed.), *The Capitol Riot: Fragile Democracy*, Greenhaven Publishing, New York, 2021.

[BLE 20] BLEAKLEY P., "Days of alt-rage: Using the Weatherman movement to deconstruct the radicalisation of the alt-right", *Contemporary Politics*, vol. 26, no. 1, pp. 106–123, 2020.

[BUE 21] BUETEL A.J., JOHNSON D., The Three Percenters: A look inside an anti-government militia, *Newlines Institute for Strategy and Policy*, available at: https://newlinesinstitute.org/far-right-extremism/the-three-percenters-a-look-inside-an-anti-government-militia/, 2021.

[CHR 20] CHRISTOU M., Is the radical right spreading the coronavirus? *Open Democracy*, available at: https://www.opendemocracy.net/en/countering-radical-right/radical-right-spreading-coronavirus/, 2020.

[CON 20] CONWAY M., "Routing the extreme right: Challenges for social media platforms", *The RUSI Journal*, vol. 165, no. 1, pp. 108–113, 2020.

[COU 22] COUNTER EXTREMISM PROJECT, Extreme left groups in the United States, The Counter Extremism Project, Berlin, 2022.

[CSI 21] CSIS, Examining extremism: The oath keepers, *Center for Strategic and International Studies*, available at: https://www.csis.org/blogs/examining-extremism/examining-extremism-oath-keepers, 2021.

[CUL 21] CULP A., "Accelerationism and the need for speed: Partisan notes on civil war", *La Deleuziana*, pp. 161–171, 2021.

[DAV 21] DAVEY J., COMERFORD M., GUH J., BALDET W., COLLIVER C., A taxonomy for the classification of post-organisational violent extremist and terrorist content, Report, Institute for Strategic Dialogue, 2021.

[DEF 21] DEFENSE ADVANCED RESEARCH PROJECTS AGENCY, Extremism and insider threat in the Department of Defense, Report, United States Department of Defense, 2021.

[DIL 22] DILANIAN K., STRICKLER L., Biggest fraud in a generation: The looting of the Covid relief plan known as PPP, *NBC News*, available at: https://www.nbcnews.com/politics/justice-department/biggest-fraud-generation-looting-covid-relief-program-known-ppp-n1279664, 28 March, 2022.

[DIX 17] DIXON T., "Art and accelerationism", *Art Monthly*, pp. 1–5, 17 February 2017.

[DON 19] DONOVAN J., LEWIS B., FRIEDBURG B., "Parallel ports. Sociotechnical change from the alt-right to alt-tech", in FIELITZ M., THURSTON N. (eds), *Post-Digital Cultures of the Far Right*, Transcript, Berlin, 2019.

[DRA 07] DRABBLE J., "From white supremacy to white power: The FBI, COINTELPRO-WHITE HATE, and the Nazification of the Ku Klux Klan in the 1970s", *American Studies*, vol. 48, no. 3, pp. 49–74, 2007.

[DRU 22] DRUKER S., Leader of Alabama chapter of Oath Keepers pleads guilty to Jan. 6 charges, *United Press International*, available at: https://www.upi.com/Top_News/US/2022/03/02/oath-keepers-alabama-chapter-leader-pleads-guilty-jan6-capitol/6381646270267, 2 March 2022.

[EAT 21] EATON P.D., TAGUBA A.M., ANDERSON S.M., "3 retired generals: The military must prepare now for a 2024 insurrection", *The Washington Post*, 17 December 2021.

[ELL 20] ELLIS E.G., The meme-fueled rise of a dangerous, far-right militia, *Wired*, available at: https://www.wired.com/story/boogaloo-movement-protests, 10 June 2020.

[EVA 20] EVANS R., WILSON J., The boogaloo movement is not what you think, *Bellingcat*, available at: https://www.bellingcat.com/news/2020/05/27/the-boogaloo-movement-is-not-what-you-think, 27 May 2020.

[FED 21] FEDERAL BUREAU OF INVESTIGATION, Adherence to QAnon conspiracy theory by some domestic violent extremists, *Martin Heinrich*, available at: https://www.heinrich.senate.gov/download/qanon-fbi-response21, 4 June 2021.

[FEU 22] FEUER A., GOLDMAN A., Oath keepers leader charged with seditious conspiracy in Jan. 6 investigation, *The New York Times*, available at: https://www.nytimes.com/2022/01/13/us/politics/oath-keepers-stewart-rhodes.html, 13 January 2022.

[FIN 20] FINKELSTEIN J., DONONUE J.K., GOLDENBURG A. et al., Covid-19, conspiracy and contagious sedition: A case study on the militia-sphere, Report, The Miller Center for Community Protection and Resilience, Rutgers University, 2020.

[FRE 20] FRENKEL S., FEUER A., A total failure: The Proud Boys now mock Trump, *The New York Times*, available at: https://www.nytimes.com/2021/01/20/technology/proud-boys-trump.html, 20 January 2020.

[FRI 20] FRIEDBERG B., The dark virality of a Hollywood blood-harvesting conspiracy, *Wired*, available at: https://www.wired.com/story/opinion-the-dark-virality-of-a-hollywood-blood-harvesting-conspiracy, 31 July 2020.

[GAL 00] GALLAHER C., "Global change, local angst: Class and the American Patriot Movement", *Environment and Planning*, vol. 18, pp. 667–691, 2000.

[GIL 15] GILL P., CORNER E., THORNTON A., CONWAY M., What are the roles of the internet in terrorism? Measuring online behaviours of convicted UK terrorists, Report, VOXPol Network of Excellence, 2015.

[GRA 13] GRAHAM C., Terrorism.com: Classifying online Islamic radicalism as a cybercrime, *The Small Wars Journal*, available at: https://smallwarsjournal.com/jrnl/art/terrorismcom-classifying-online-islamic-radicalism-as-a-cybercrime, 2013.

[GRE 21] GREEN E., The conservatives dreading – and preparing for – Civil War, *The Atlantic*, available at: https://www.theatlantic.com/politics/archive/2021/10/claremont-ryan-williams-trump/620252/, 1 October 2021.

[GUH 21] GUHL J., DAVEY J., A safe space to hate: White supremacist mobilisation on Telegram, *ISD*, available at: https://www.isdglobal.org/isd-publications/a-safe-space-to-hate-white-supremacist-mobilisation-on-telegram/, 2021.

[HAT 18] HATEWATCH STAFF, Are white nationalists turning on Trump?, *Southern Poverty Law Center*, available at: https://www.splcenter.org/hatewatch/2018/11/27/are-white-nationalists-turning-trump, 27 November 2018.

[HEA 11] HEAD M., *Crimes Against the State: From Treason to Terrorism*, Routledge, Oxon, 2011.

[HOF 20] HOFFMAN B., CLARKE C., The next American terrorist, *The Cipher Brief*, available at: https://www.thecipherbrief.com/article/united-states/the-next-american-terrorist, July 2, 2020.

[HUN 20] HUNTER L.Y., GRIFFITH C.E., WARREN T., "Internet connectivity and domestic terrorism in democracies", *International Journal of Sociology*, vol. 50, no. 3, pp. 201–219, 2020.

[ISD 21] ISD, Between conspiracy and extremism: A long Covid threat?, Paper, Institute for Strategic Dialogue, Amman, 2021.

[JAR 22] JARED H., After the insurrection: How domestic extremists adapted and evolved after the January 6 US Capitol attack, *The Atlantic Council*, available at: https://www.atlanticcouncil.org/in-depth-research-reports/report/after-the-insurrection-how-domestic-extremists-adapted-and-evolved-after-the-january-6-us-capitol-attack/, 2022.

[JOH 20] JOHNSON B., How terrorists are trying to make coronavirus more friend than foe, *Homeland Security Today*, available at: https://www.hstoday.us/subject-matter-areas/wmd/page/4/, 14 April 2020.

[JON 20] JONES S.G., DOXSEE C., HARRINGTON N., The escalating terrorism problem in the United States, CSIS Policy Briefs, Center for Strategic and International Studies, 2020.

[JON 21] JONES S.G., DOXSEE C., Examining extremism: Antifa, *Center for Strategic and International Studies*, available at: https://www.csis.org/blogs/examining-extremism/examining-extremism-antifa, 2021.

[KEN 21] KENES B., Boogaloo Bois: Violent anti-establishment extremists in festive Hawaiian shirts, *European Center for Populism Studies*, available at: https://www.populismstudies.org/boogaloo-bois-violent-anti-establishment-extremists-in-festive-hawaiian-shirts/, 2021.

[KHA 21] KHALIL L., GNET survey on the role of technology in violent extremism and the state of research community-tech industry engagement, Report, International Centre for the Study of Radicalisation, King's College, London, 2021.

[KIL 21] KILCULLEN D., American mythology dies on the Hill, *The Australian*, available at: https://www.theaustralian.com.au/commentary/american-mythology-dies-on-thehill/newsstory/f0b8e43c8e1fef89d79c5d6632c7466d, 7 January 2021.

[KRI 21] KRINER M., LEWIS J., "The evolution of the Boogaloo movement", *CTC Sentinel*, vol. 14, no. 2, pp. 22–32, 2021.

[KUT 20] KUTNER S., Swiping right: The allure of hyper masculinity and cryptofascism for men who join the Proud Boys, ICCT Research Paper, International Centre for Counter-Terrorism, The Hague, 2020.

[LAU 17] LAURENCE M.R., "Speed the collapse?: Using Marx to rethink the politics of accelerationism", *Theory & Event*, vol. 20, no. 2, pp. 409–425, 2017.

[LYO 17] LYONS M.N., Ctrl-Alt-Delete: The origins and ideology of the alternative right, Report, Political Research Associates, Somerville, MA, 2017.

[LYT 21] LYTVYNENKO J., HENSLEY-CLANCY M., The rioters who took over the Capitol have been planning online in the open for weeks, *BuzzFeed News*, available at: https://www.buzzfeednews.com/article/janelytvynenko/trump-rioters-planned-online, 6 January 2021.

[MAI 18] MAIN T.J., *The Rise of the Alt-Right*, Brookings Institution Press, Washington, DC, 2018.

[MAR 20] MARION N.E., TWEDE J., *Cybercrime: An Encyclopedia of Digital Crime*, ABC-CLIO, Santa Barbara, CA, 2020.

[MCC 21] MCCLAIN C., VOGELS E.A., PERRIN A. et al., The Internet and the Pandemic, *The Pew Research Center*, available at: https://www.pewresearch.org/internet/2021/09/01/the-internet-and-the-pandemic/, 1 September 2021.

[MCS 21] MCSWINEY J., JASER G., HAMMER D., Alt-tech and online organising after the Capitol riots, *Global Network on Extremism and Technology*, available at: https://gnet-research.org/2021/01/21/alt-tech-and-online-organising-after-the-capitol-riots/, 21 January 2021.

[MEL 21] MELLEY T., BUTTER M., KNIGHT P., Conspiracy theory in American narrative, in BUTTER M. and KNIGHT P. (eds), *Routledge Handbook of Conspiracy Theories*, Routledge, Abingdon, VA, 2021.

[MIL 20] MILLER C., "There is no political solution": Accelerationism in the White Power Movement, *Southern Poverty Law Center*, available at: https://www.splcenter.org/hatewatch/2020/06/23/there-no-political-solution-accelerationism-white-power-movement, 23 June 2020.

[NAT 21] NATIONAL THREAT INTELLIGENCE CONSORTIUM, Domestic terrorism: A persistent threat in 2021, Report, District of Columbia Homeland Security and Emergency Management Agency, 2021.

[NOY 22] NOYS B., "Accelerationism, Brexit and the problem of 'Englishness'", *Arguments Within English Theory*, vol. 32, no. 5/6, pp. 586–592, 2022.

[NSC 21] NSC, National strategy for countering domestic terrorism, Report, National Security Council, Washington, DC, 2021.

[OBR 17] O'BRIEN L., MATHIAS C., The maniac neo-Nazis keeping Charles Manson's race war alive, *The Huffington Post*, available at: https://www.huffpost.com/entry/alt-right-charles-manson-atomwaffen_n_5a146921e4b03dec824892e6, 21 November 2017.

[ONG 20] ONG K., "Ideological convergence in the extreme right", *Counter Terrorist Trends and Analyses, International Centre for Political Violence and Terrorism Research*, vol. 12, no. 5, pp. 1–8, 2020.

[PUB 22] PUBLIC SAFETY CANADA, Currently listed entities, *Government of Canada*, available at: https://www.publicsafety.gc.ca/cnt/ntnl-scrt/cntr-trrrsm/lstd-ntts/crrnt-lstd-ntts-en.aspx, 2022.

[ROC 17] ROCCA N.R., Mobilization and radicalization through persuasion: Manipulative techniques in ISIS' propaganda, Paper, University of Coimbra, 2017.

[ROT 21] ROTHSCHILD M., *The Storm is Upon Us: How QAnon Became a Movement, Cult, and Conspiracy Theory of Everything*, Melville House Publishing, Brooklyn, NY, 2021.

[SAV 19] SAVAGE R., "Populism in the U.S.", in DE LA TORRE C. (ed.), *Routledge Book of Global Populism*, Routledge, New York, 2019.

[SPA 00] SPARK A., "Conjuring order: The New World Order and conspiracy theories of globalization", *The Sociological Review*, vol. 48, no. 2, pp. 46–62, 2000.

[STA 11] STANTON Z., The problem isn't just one insurrection. It's mass radicalization, *Politico*, available at: https://www.politico.com/news/magazine/2021/02/11/mass-radicalization-trump-insurrection-468746, 11 February 2011.

[STA 21] STANO S., "The internet and the spread of conspiracy content", in BUTTER M., KNIGHT P. (eds), *Routledge Handbook of Conspiracy Theories*, Routledge, Abingdon, VA, 2021.

[USC 18] USCINSKI J.E., DEWITT D., ATKINSON M.D., "A web of conspiracy? Internet and conspiracy theory", in DYRENDAL A., ROBERTSON, D.G., ASPREM E. (eds), *Handbook of Conspiracy Theory and Contemporary Religion*, Brill, Leiden, 2018.

[VEL 21] VELÁSQUEZ N., MANRIQUE P., SEAR R. et al., "Hidden order across online extremist movements can be disrupted by nudging collective chemistry", *Nature Scientific Reports*, vol. 11, no. 9965, 2021.

[WAL 07] WALL D.S., "Policing cybercrimes: Situating the public police in networks of security within cyberspace", *Police Practice and Research*, vol. 8, no. 2, pp. 183–205, 2007.

[WAL 20] WALKER H., WINTER J., White supremacists discussed using coronavirus as a bioweapon, *Huffington Post*, available at: https://www.huffpost.com/entry/white-supremacists-coronavirus-bioweapon_n_5e76a0ebc5b6f5b7c5458af2, 21 March 2020.

[WAT 21] WATSON E., Defense Secretary nominee vows to root out enemies who lie within our own ranks, *CBS News*, available at: https://www.cbsnews.com/news/lloyd-austin-defense-secretary-confirmation-watch-live-stream-today-01-19-2021/, 19 January 2021.

[WIL 20] WILLIAMS T., GOLDMAN A., MACFARQUHAR N., Virginia capital on edge as F.B.I. arrests suspected neo-Nazis before gun rally, *The New York Times*, available at: https://www.nytimes.com/2020/01/16/us/fbi-arrest-virginia-gun-rally.html, 21 January 2020.

[YAT 16] YATES E., LAFREE G., JENSEN M., KANE S., Profiles of individual radicalization in the United States (PIRUS), *National Consortium for the Study of Terrorism and Responses to Terrorism*, available at: https://www.start.umd.edu/data-tools/profiles-individual-radicalization-united-states-pirus, 2016.

[ZOO 96] ZOOK M. "The unorganized militia network: Conspiracies, computers, and community", *Berkeley Planning Journal*, vol. 11, pp. 26–48, 1996.

4

Cybercrime in Brazil After the Covid-19 Global Crisis: An Assessment of the Policies Concerning International Cooperation for Investigations and Prosecutions

4.1. Introduction: Brazilian cybercrime and the Covid crisis impact

Cybercrime is not a new subject for Brazilian authorities. A 2015 article states that Brazil was the second most vulnerable country to cybercrime (Muggah and Thompson 2015). The article also informs us that Brazilian society has lost between 4.1 and 4.7 billion US dollars in data theft and financial fraud. Most of the targets of cybercrime were businesses and citizens who had money in banks. The article concludes that the Brazilian government and society should implement more reforms to tackle cybercrime. But how is the country performing reforms to achieve this? This chapter answers this question with two evaluations: the most recent reforms and what else is coming shortly. In addition, it traces the current situation of cybercrime in Brazil after the pandemic crisis using multi-sources data. It uses quantitative data available from SaferNet and the Federal Prosecutors' Office of Brazil and qualitative data from four exploratory interviews with law enforcement personnel (LEP). SaferNet is a non-profit and non-governmental organization that dates back to 2005 (SaferNet, 2021a). Its

---

Chapter written by Alexandre VERONESE and Bruno CALABRICH.

mission is to support a safer usage of the Internet. It has partnerships with many Brazilian state agencies, enterprises and other non-governmental organizations (SaferNet, 2021b). The Federal Prosecutors' Office of Brazil is responsible for fighting cybercrime when it has transnational links. The LEP interviews serve two purposes. First, they all have a great deal of experience in cybercrime and international cooperation for criminal matters and, therefore, have a lot to say about the subject. Secondly, they show how the Covid crisis was a watershed moment for cybercrime fighting in Brazil, as LEP-2 describes:

> We had a rise with cybercrime. (…). Many people started using the Internet because of the pandemic, and they do not have the skills to protect themselves. They become easy prey. (...). In my perception, there was a rise both in financial cybercrime and in the sexual abuse of minors.

The increasing expansion of cybercrime led the LEP to organize themselves into units and groups all over the country. More pressure from cybercrime led to an improvement in the authorities' response. The most visible change was upgrading the debate about international cooperation to fight against cybercrime. This debate led to two major reforms. The first reform was the passing into law of some bills to better proscribe some conducts as crimes. The second reform is ongoing. It is the process of better integrating Brazil into the international networks against cybercrime. We use data to provide not only a description of the Brazilian case but for interpreting it to raise a theoretical legal problem concerning the interaction between authorities of many different countries to fight against cybercrime. The conceptual transposition of cybersecurity to the social sciences is an ongoing debate that requires more empirical research to build up an epistemological debate (Loiseau 2020).

This chapter has six main sections and a conclusion. Section 4.2 presents a literature review to enlighten the cybercrime subject, extracting some elements from them to assess the Brazilian case. Section 4.3 presents a theoretical debate. We consider that fighting against cybercrime demands national efforts to improve international cooperation. Nonetheless, that will require an increase in both legal and technical solutions. The legal community must try to improve a specific legal concept to supply the basis for cooperation, for one side. It will also be relevant to create more technological systems to exchange and transfer data between authorities securely. From a legal point of view, our theoretical suggestion is to use the concept of legal interoperability to solve the issue. This concept is better than legal equivalency or the national adequacy level of protection because it comes from the ongoing governance and management debate, intertwining with technical standards discussions. Section 4.4 describes the data about cybercrime in Brazil, focusing both on SaferNet and the Federal Prosecutor's Office databases. It shows the significant increase in cybercrime in Brazil, blended with some information from the interviews of the LEP. This section will also provide an opportunity to understand the current situation of the Brazilian legal system on this subject and how the LEP is pushing forward better to connect the country to the international networks against cybercrime.

The most important fact is that Brazil finally is in the ongoing process of acceding to the Budapest Convention. The Brazilian Senate promulgated Congress Decree No. 37 and published it on December 17, 2021 (Brazilian Federal Senate, 2021). The last step is the deposit of the ratification. We can assume that the Covid crisis was fundamental for the country to pursue accession to this Convention. The next section will describe the Brazilian capacity to fight cybercrime, relying on data from an Organization of American States (OAS) report. It will also describe how the Brazilian legal system can integrate itself with the Budapest Convention. The LEP interviews provide information about how Brazil made this accession and how they perceived that this was a water-shedding moment. None of the interviews mention the BRICS eThekwini Declaration of 2013. This declaration is part of a major effort for engagement between Brazil, Russia, India, China and South Africa on various issues. One topic was infrastructure telecommunications and Internet security (BRICS Information Center, 2013). Section 4.6 describes and interprets how the Brazilian LEP had to establish cooperating methods without having the best legal tools to hand. We will transcribe and interpret LEP interviews to show how they had to use some old international cooperation channels, like Interpol.

Moreover, they talk about alternative ways of cooperating without a national framework with full integration of the international means. This cooperation will pave the way to a return to the theoretical problem. Section 4.7 describes a crucial subject: the problem of establishing an institutional basis for the country to cooperate in the global fight against cybercrime. Also, it brings forth two legal and social issues about international cooperation against cybercrime and resumes the theoretical debate. The first issue is the compatibility of criminal systems. The most notable example comes from United States law, which does not consider some speeches and discourses to be felonies or misdemeanors. The US First Amendment of the Constitution provides a more flexible basis to interpret the freedom of speech protection. The United States' constitutional tradition is far more flexible than European nations' legal systems. Such flexibility is also higher in Latin America and even the United Kingdom. The second issue is the actual compatibility of procedures, complicating international cooperation. This issue also has ties with the debate about the extraterritorial application of the law. The chapter concludes that an increase in legal and technical solutions will come in the next few years after applying efforts from the authorities with a suitable theoretical model. The international legal community must still try to develop a specific legal concept to supply the basis for cooperation.

4.2. Cybercrime in the literature and the Brazilian case

Despite the vast future challenges, we can assess that some improvements have happened in the international arena to fight against cybercrime. Peter Grabosky wrote a paper almost 20 years ago to describe the panorama of both Australia and the international issues about cybercrime (Grabosky 2003). During the beginning of the 21st century, the scenario was very different. Since then, many countries have passed significant reforms into law. Also, nowadays, there are some international standards. Therefore, the problems evolve to more practical issues like defining tools aiming for effective results. The Brazilian case is exemplary. In this section, we will provide a literature review that will be helpful to understand cybercrime in general.

The legal debate about cybercrime between the 1980s and 1990s focused on hacking as a primary subject. In the United States, hacking federal systems became a crime under Section 1030 of Title 18 (Crimes and Criminal Procedure, and Appendix) of the US Code. Also, many states followed this track and have statutes' provisions qualifying hacking as a crime. One example is that the US Congress approved a bill passed into law in 1984: the Counterfeit Access Device and Computer Fraud and Abuse Act (GovInfo 1984). The United States federal statute dates from the same year that Steven Levy published his book *Hackers: Heroes of the Computer Revolution* (Levy 1984). Some years after, Bruce Sterling published *The*

Hacker Crackdown: Law and Disorder on the Electronic Frontier (Sterling 1992). At that time, the Internet was far from what it is nowadays.

Nevertheless, these visionaries saw something coming. This subject was coming to stay both in the public debate and in the legal literature. Harry Rubin, Leigh Fraser, and Monica Smith wrote an article about the criminal potential over the Internet in 1995 (Rubin et al. 1995). They stated that the adaptation of the law to the Internet scenario was something hard to do and that even the 1984 (Counterfeit Access Device and Computer Fraud and Abuse Act) and the 1986 (Electronic Communications Privacy Act) federal statutory amendments were unable to solve most of the legal issues. For Jo-Ann M. Adams, the solution would be to amend some more of the 1984 Act, passing into law some other provisions, as in the case of the Communications Decency Act (CDA) of 1995, to effectively rule over the "anarchy on the Internet" (Adams 1996).

Defining clear lines to separate common crimes from computer crimes was still very difficult for criminal law. One reason was that some criminal conducts involve intellectual property and others intangible goods. David Wall tried to classify cybercrime in parallel to actual crimes. He wrote about "cyber-trespass", "cyber-theft", "cyber-obscenity" and "cyber-violence" (Wall 1998, 1999). His objective was to criticize both authors focused on the crimes themselves and authors who tried to classify "new" social behaviors. Nonetheless, his classification at that time was too wide and was lacking some refinement. Another way to classify criminal activity on the Internet was to look at the hardware involved. Carl Benson, Andrew Jablon, Paul Kaplan, and Mara Elena Rosenthal took this approach. They define three possibilities to classify "computer crime" (Benson et al. 1997). Computers could be targets of criminal activities as someone could steal them. Also, a computer could be the place of a crime. Moreover, finally, they could be instruments of traditional crimes. We will see that this definition has become more refined over the years. However, at some point, the logical basis of this classification is still valid. Also, the expansion of the Internet comes along with the international aspect of cybercrime.

One example is figuring out how to prosecute criminals outside the borders perpetrating intellectual property crimes (CHR 97, ZEV 97–98). In addition, the problem of training law enforcement personnel to pursue and prosecute cybercrime was an issue in the past and still is a problem. Some saw technical solutions, and some even saw the possibility of using private companies to help law enforcement (Mitchell and Banker 1998). The academic debate and policy efforts against cybercrime generated some international solutions. The most notable is the Council of Europe's Convention on Cybercrime of 2001.

Nowadays, it is not difficult to see that the first propositions to classify cybercrime are no longer a problem. As Bilel Benbouzid and Daniel Ventre describe, relying on David Wall's most recent work, there are three kinds of cybercrime (Wall 2007; Benbouzid 2016). First, the law framed those crimes before the computer age (cyber-assisted crimes). Most property and document frauds fall under this definition. Forgery of a credit card would still be a crime, whether using the Internet is part of the criminal activity or not. The second kind encompasses crimes that the Internet makes easier and plays an essential role (in cyber-enabled crimes), like the diffusion of computer viruses. The third category would be impossible to perform without the Internet, like attacking servers (cyber-dependent crime). From a legal point of view, it is essential to explain that social conduct needs a clear description to be a felony or a misdemeanor legally. Otherwise, the widespread analogical application of criminal law would be erratic and even unfair.

It is crucial to indicate that this chapter follows the effort to provide data to a still ongoing research program. The article by Bilel Benbouzid and Daniel Ventre describes four subjects in the sociology of cybercrime (Benbouzid and Ventre 2016). The first is the mapping of criminal hackers. The second is having more significant surveys about the crimes and their victims. The third is the qualitative analysis of Internet communities and deviant behavior. The fourth is a sociological analysis of the ongoing transformation of Internet policing, which demands research about surveillance. This chapter's contribution falls under the fourth subject. It shows how Brazilian law enforcement personnel is actively transforming the national system to integrate it into a global network against cybercrime. It is also important to mention that the Covid crisis pushed the Brazilian law enforcement system to bolden its capacity to fight against cybercrime, as LEP-1 explains. Brazil created a payment for people affected by the crisis, as the United States administration did (Economic Impact Payments, managed by the Internal Revenue Service – IRS):

> With the pandemic, we saw problems involving the federal government's emergency payment [Coronavirus aid]. In those cases, we saw specific crimes directly involving the Internet.

Criminals used fake data and Internet cashing systems to receive payments they would not qualify for. We saw criminals using dead people's data to cash checks. Also, we saw criminals stealing data from persons who did not need the payments and altering them just to cash.

These cases gave the law enforcement personnel a lot of work. A critical part of the process had skilled staff identifying frauds and criminals and recovering funds (Lima 2020). The criminals were primarily Brazilians. Nevertheless, they could use a lot of different routes to transfer funds abroad. A tiny part of the federal money was recovered (Rodrigues 2020). This lack of success relates to the need for international cooperation to tackle contemporary cybercrime. The following section will present a model to introduce the subject of international cooperation from a theoretical point of view.

4.3. A theoretical model for international cooperation

It is possible to provide a model to compare the countries' readiness for cybercrime fighting and provide international cooperation procedures and measures. We will use the concept of interoperability to reach an advanced theoretical model. Nonetheless, we can again state that it is impossible to understand interoperability in legal and social terms in the same manner as technical systems do (Belli and Foditsch 2016). Legal interoperability can happen in various degrees, while technical interoperability occurs with a binary response (yes or no). The first variable set of our model focuses on the national jurisdictions. But other variables can be added to improve the complexity of any analysis, like regional integration.

Focusing on national level legal systems, on the one hand, we can evaluate the crimes' definitions compared to some standard social conducts abroad. On the other hand, we can evaluate the procedural practices to infer how they are like some standards abroad. The second set focuses mainly on international means and channels. We can begin evaluating how well one country can provide aid to other jurisdictions. The model can also measure the country's capacity to conduct local investigations and prosecutions, requesting international cooperation by providing data and distributing them along with these four variables.

Nevertheless, we will criticize two alternative concepts before drafting this model in an image. The first concept is compatibility. The building of computer and technical systems can have open and closed (proprietary) sources and standards. The open-source idea is to enable the software or hardware to have no limits on their adaption to fit almost any system. The notion of compatibility here provides a powerful image. It does not matter what the operating system is, as long it is open, one can adapt it to fit. The openness of the source code, or technical standard, will allow the new module to be adapted to the main one and vice-versa. Unfortunately, this powerful image cannot have a precise translation to the understating of legal systems. Legal systems are social constructs that have a high degree of ruling behavior, which is more unstable than technical constructs. This social behavior changes over time and varies from one place to another. Therefore, evaluating or measuring countries' legal compatibility is unattainable theoretically. We can use this concept as a metaphor. But it will be useless for our theoretical purposes. The other concept is adequacy, which is building metrics to measure data protection levels. This concept is still under construction. From a legal-formalist point of view, both the General Data Protection Regulation and the Directive (EU) 2016/680 use this concept. These two European Union legal documents provide guidelines to evaluate how the European Commission can determine whether a country is adequate to receive and provide personal data to member states. Figure 4.1 below summarizes the concept of adequacy, relying on Article 36 of the Directive (EU) 2016/680.

Notwithstanding having the above framework, two rulings – known as Schrems I and II – from the Court of Justice of the European Union brought down the European Commission's decisions that recognized the United States as an adequate country (Veronese 2021). Do those rulings impose the halting of dataflows between the European Union member states and the United States? No. The decisions recognized that the flows might lawfully happen, relying on a different data protection basis, like standard contractual clauses. The concept of adequacy will evolve in the next few years. So, how can we think differently about international cooperation from a theoretical point of view? The concept of legal interoperability appears in the literature as a concept that has its origin in the management debate (Morando 2013; Mellone 2014), even in discussing e-justice (Küster 2012). Also, this concept still serves as a basis to build public policies, like fostering the interoperability of national systems to provide social rights among the member states of the European Union (European Union 2017).

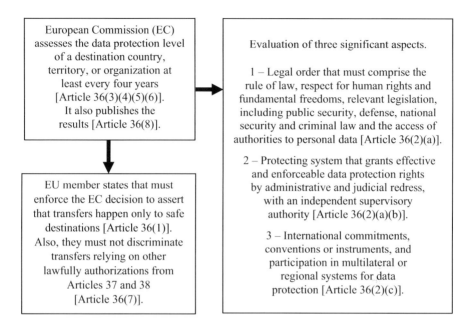

Figure 4.1. *Article 36 criteria for the European Commission to assess the data protection level of adequacy of a non-EU member country, international organization. Source: authors' design*

Amadeo Santosuosso and Alessandra Malerba explain legal interoperability as an alternative concept for having a new glaze on comparative law and international law (Santosuosoo and Malerba 2014). The central aspect of the concept is its offer to look at dynamic interchanges between different legal contexts. They also highlight that the theoretical center of the debate lies in both linguistic and cultural understandings. They define three main situations. There is a sole legal system with just one language in the first situation. It is the temporal transformation by which societies shape new legal concepts or modify their interpretation. In the second situation, different legal systems use the same language, like the United States and the United Kingdom. In the third situation, there are different legal systems with different languages. One could state that the core solution is to have a perfect translation. However, that is precisely the problem. It is almost impossible to have it. So, the solution would be to have an equivalent solution. This model of legal interoperability could lead to a better understanding of differences:

It seems to us that those traditional comparative disciplines (which indeed focus on similarities and differences between two or three legal systems but, in doing so, mainly tries to lead to uniformity and propose standardized answers to legal problems) are not able to face all this massive quantity of differences among so many legal systems. Instead, legal interoperability: (a) focuses on differences rather than on similarities; (b) has as its primary goal putting in contact (and making operative) elements that naturally would be separated because of some conceptual or linguistic misalignment (beyond the pure false/true opposition) (…); and (c) offers a vision of more than two legal particles/systems working together (Santosuosso and Malerba 2014).

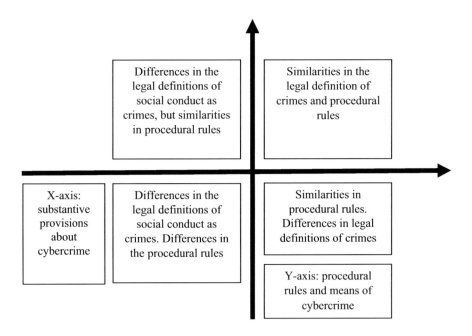

Figure 4.2. *Classification of national systems for legal interoperability assessment purposes. Source: authors' design*

It is interesting to produce an image to debate the subject of international cooperation on criminal matters. It will not surprise sociologists that

Figure 4.2 is like Max Weber's ideal formal and substantial law classification (Trubek 1972; Swedberg 1998). We can imagine two national legal systems whose prescriptions and sociolegal practices are interoperable. In addition, we can think about a scenario in which two legal systems are utterly not interoperable. In between those extreme cases, we see the real scenarios. We could use this model to understand how fragments of a legal system may find application in another, producing effective results.

The crucial point of the model is to recognize differences in various legal concepts from many legal systems to promote the possibility of adapting existing ones or creating new ones. That construction is far from easy. At first glance, from a practical point of view, one can see this as an operational failure. Nevertheless, if the government and researchers evaluate those "cooperation failures" analytically, they may use them to propose solutions. A conclusion is that the best solution is not imagining the theoretical model of legal interoperability as a means for harmonization or standardization, as Rolf H. Weber describes (Weber 2014). Many technical international forums dedicate themselves to doing this. The focus must be on adapting legal solutions, always remembering that those legal concepts and systems are social, cultural, and linguistic products. They are very different from technical rules.

4.4. The evolution of cybercrime in Brazil

Brazil has gone through a significant increase in cybercrime against electronic systems and frauds using networked computers. In recent years, public opinion has become aware of the threats through two cases. The first was a ransomware attack against the High Court of Justice (*Superior Tribunal de Justiça*, in Portuguese), whose jurisdiction encompasses special appeals to reverse federal and state appellate courts' rulings. To undertake this kind of attack, the hackers infiltrate the digital systems and then cipher all the stored data on the servers with high-grade algorithms. This kind of attack is becoming more common, and many enterprises and public institutions are falling prey to them, as LEP-2 explains in his interview:

> We are aware of the most recent threats. Among them, the uttermost serious was the hacking of many Brazilian courts. It began with the well-known High Court of Justice case. However, hacking happened also against the federal judiciary

system when hackers altered even legal opinions. During the pandemic, the online systems of the government and enterprises became more vulnerable because people were working from their personal computers at home. (…). The ransomware cases are rising. They make cyber extortion demanding money for data encrypted by them. We saw this crime skyrocket. Furthermore, they started attacking public agencies and courts after the pandemic, something new. We think that intensive online work is the cause. (…). The crisis pushed out a significant portion of the people to work from home, and this happened without preventive measures. It was all of a sudden.

The High Court of Justice reacted promptly by mobilizing its information technology team to block unlawful access to the systems and restore the data from backups. Nevertheless, these systems' restoration takes some time. One can measure the social damage of the attack by assessing an estimate of delays to rule appeals and lawsuits. The court only ruled on extraordinary measures using on-premises software during the restoration period. The daily routine suffered a break. The second case was a hacker attack against federal authorities' mobile phones. The Brazilian federal police started an investigation called "spoofing", whose name refers to the technique of intrusion. The hackers gain unlawful access to digital systems by pretending to use legitimate credentials in these attacks. The hackers had access to all the communications from public officials stored on their phones and to data from third parties. These cases are well known because they attack the state organizations or their officials. However, many citizens and private enterprises are falling victim to similar attacks. One digital security firm, Karpersy LLC, estimates that Brazil is experiencing an increase of 300% of violations throughout the Internet (Rolfini 2020). Also, the same firm reported in 2021 that Brazil is leading the region in the number of incidents.

A preliminary basis for comparison is on the SaferNet website (SaferNet, *Indicadores* 2021). It receives anonymous complaints about illegal content on websites. SaferNet has two channels. The first is a hotline, a website that collects complaints. The second is a helpline that provides humanitarian assistance, talking or chatting. In the past, SaferNet aggregated the data from two other channels: the Brazilian federal police and the Brazilian national secretariat of human rights, as one can see using the WayBackMachine that stores old data from various websites (Internet Archives 2022). Today, the SaferNet website relies on data only from its own channels. This project covers receiving complaints and evaluating the takedown measures. The domains with the most content takedowns during almost 15 years of service (2006–2021) are Orkut, Facebook, YouTube, Twitter, Instagram, XVideos, UOL and imgsrc.ru. Table 4.1 shows a comparison between complaints and removals.

Domain	Description	Complaints	Removal
orkut.com.br	A social network that was very popular in Brazil until 2014.	334,231	334,231
facebook.com	One of the more extensive social networks nowadays.	82,989	29,970
youtube.com (Google)	The biggest video streaming platform nowadays.	15,427	13,509
twitter.com	One of the more extensive social networks nowadays.	35,849	12,597
instagram.com (Facebook)	One of the more extensive photo-sharing networks nowadays.	12,010	11,948
xvideos.com	A pornographic website that allows content uploading.	11,822	8,180
uol.com.br	A Brazilian content provider.	4,495	3,972
imgsrc.ru	A Russian images provider that stocks mostly pornographic content.	3,907	3,092

Table 4.1. *Domains with most complaints to SaferNet from 2006 to mid-2021. Source: SaferNet (2021), after authors' treatment. Data were collected on 30 October 2021*

A significant effort from law enforcement is taking down unlawful content. Nonetheless, most of the removals require collaboration by Internet service providers. The table shows that some Internet content providers are more responsive than others. Further in this chapter, we will describe the

Yahoo versus LICRA case, which is very illustrative of this issue. The providers may rely on their nation-state legal system to evaluate a request for a takedown, as LEP-1 describes:

> When we make the takedown request, you can see that they perform a legal analysis to identify whether the content is illegal. (...). If you cannot prove this, you receive an answer denying the request. The providers explain that the content may be within the freedom of expression. (...). We have the impression that the significant legal orientation comes from abroad.

From the beginning, taking down content from the Internet has been a significant issue. The option of having a non-governmental organization to assist authorities probably helps a smoother relationship between the Big Tech companies and law enforcement agencies. Table 4.1 shows the successful effort of the partnership between SaferNet and the Brazilian authorities over almost 15 years. Despite acting as a channel, SaferNet also categorizes the complaints by several kinds of crimes. All of them are both in the Brazilian Criminal Code and in international treaties. Brazil has signed and ratified to accede to three applicable United Nations treaties: the Convention on the rights of the child and its optional protocol on the sale of children, child prostitution, and child pornography (United Nations, 2000a); the International Convention on the elimination of all forms of racial discrimination (United Nations 1965); and the Protocol to Prevent, suppress and punish trafficking in persons – especially women and children, supplementing the United Nations Convention against transnational organized crime (United Nations, 2000b). Table 4.2 describes the crimes categorized by SaferNet. In the first draft of this research, the data collected aggregated two more sources. Nonetheless, SaferNet changed its website to inform just numbers directly from their channels (hotline and helpline) (Internet Archives 2022). So, Table 4.2 was recreated to reflect this modification.

Table 4.2 shows a lot about how Brazilian users use the Internet for some unlawful activities. We can see that posting racist content experienced a downturn from 2018 to 2019, then experiencing a major increase in 2020. The diffusion of child pornography and neo-Nazi content had a boom in 2020 and 2021. The only hypothesis available to those changes may come from the political environment or the country's historical roots (Oliveira and Almeida 2014; Trindade 2018; Knoth 2022). Comparing the numbers for

2019 with 2020, we can see that only "religious intolerance" content experienced some kind of downturn. All other unlawful types of content rose to some degree. However, Brazil is experiencing a rise in cybercrime, not only measurable by unlawful content posting. Another informant, LEP-3, mentions both types of crimes (posting unlawful content and property crimes, like theft or fraud):

Sorts of crimes	2017	2018	2019	2020	2021
Mistreating animals	4,821	1,142	885	4,184	3,081
Promoting phobia or prejudice against lesbians, gays, bisexuals and transgender people	2,591	4,244	2,752	5,293	5,347
Promoting neo-Nazism	1,172	4,244	1,071	9,004	14,476
Child pornography	33,909	60,008	48,576	98,244	101,833
Religious intolerance	1,459	1,084	1,413	1,321	759
Promoting or performing xenophobia	1,395	9,703	978	2,066	1,097
Promoting or performing racism	6,166	8,336	4,310	10,684	6,888
Performing or promoting violence and discrimination against women	961	16,717	7,112	12,698	8,174
Human trafficking	612	509	392	459	326
Promoting and urging for crimes against peoples' lives	10,611	27,713	8,182	11,852	7,390
Total	63,697	133,700	75,671	155,805	149,371

Table 4.2. *Categories of complaints to SaferNet demanding content removal (2017–2021). Source: SaferNet (2022), after authors' treatment*

> Our cybercrime group works on legal cases at the court of appeals. (...). We mostly see online frauds, online property crimes, and racist content. All crimes that use the Internet and are international fall under federal jurisdiction because they are in international treaties. For example, the crimes of child abuse come from the Optional Protocol of 2000 of the United Nations Convention on the child's rights, which dates from 1989. (...). Another example comes from racism, which fell under federal jurisdiction when Brazil acceded to the International Convention on eliminating all forms of racial discrimination in 1965. When common crimes intertwine with money laundering, they also fall under federal jurisdiction. We work basically against those crimes.

The rise of hateful speech and content is a contemporary issue. If we aggregate the data from SaferNet, the year 2020 provides a very concerning image. 2020 is the year when the Covid crisis had its first peak. Therefore, we can infer that the global health crisis settled the ground for the rise of hate speech and content diffusion. As SaferNet keeps track of those crimes and helps authorities, it became a target in 2021 (Impreza 2021). Its founder and primary manager fled from Brazil to Germany because he was under threat by criminals. In addition, someone kidnapped an employee from the non-governmental organization, and a family member of its founder suffered an assault. This situation also affects police staff, judges and prosecutors, who can become targets of crimes in retaliation for investigations and prosecutions. SaferNet is still online and providing its tools. The Federal Prosecutors Office keeps track of many lawsuits and investigations. It has prepared a manual, primarily because of the crisis, for law enforcement officers in general, as LEP-3 describes:

> Cybercrime has a lot to do with the Covid crisis. We have a guide for Internet investigation and fighting misinformation about the pandemic. In this guide, we cover not only misinformation but also fraud. We cover everything that came along with the Covid crisis because the numbers rose unbelievably.

It is possible to see the impact of the Covid crisis by looking at the data collected from the Federal Prosecutor's Office computer system. Figure 4.3 shows three categories of crimes under federal jurisdiction (electronic intrusions in federal systems, crimes against children and racial crimes). Also, the figure shows investigations by the Brazilian federal police and actual prosecutions, that is, the judicial process, which has finished its investigatory phase.

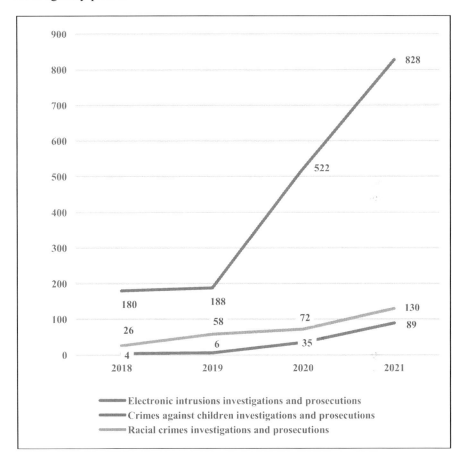

Figure 4.3. *Federal procedures against cybercrime by law enforcement personnel. Source: Federal Prosecutor's Office of Brazil (Federal General Prosecutors' Office of Brazil, 2022), after authors' treatment, aggregating data*

The type of cybercrime most investigated and prosecuted by the federal law enforcement is that against children. As LEP-3 said: "The most relevant field is child protection, despite racism being a major issue in Brazil. The protection of children is a priority". One can interpret that LEP is responsive to social claims. Remember that we have many complaints pursuing the takedown of children's pornography content through SaferNet (Table 4.2). One can only infer that we are talking about the most socially perceived cybercrime during the Covid crisis. As LEP-3 said, Brazilian law enforcement gave priority to fighting against this. Nonetheless, when looking at other lines of investigation and prosecution, it is possible to see that law enforcement procedures against electronic intrusion cybercrime more than doubled during the pandemic. Also, the investigations and prosecutions almost doubled when we look at racial cybercrime. The data indicate that the Covid crisis impacted the number of cybercrimes. However, the storm also came along with some winds of change. The following section will describe the status of the Brazilian legal system regarding cybercrime. It will show that the Brazilian LEP is making a significant effort to deeply integrate the country into international cooperation systems. The most unmistakable evidence is the accession to the Budapest Convention. Nonetheless, some more actions are ongoing, and we will describe them.

4.5. The evolution of the Brazilian legal system concerning cybercrime and its connection to the international regime

One critical report assesses Brazil's institutional capacities to fight against cybercrime. The Global Cyber Security Capacity Centre produced a report for the Organization of American States (OAS) in 2020. It covers criminal activities through computers and crimes against digital systems (Global Cyber Security Capacity Centre 2020). This research dates from between 2018 and 2020, and it deals with a wide array of issues. One of them is especially important to this chapter: the readiness of the countries of the continent to answer formal and informal demands of international collaboration against cybercrime. This subject is in Section D4.3 of the report. It has the following title: "Institutional frameworks for formal and informal collaboration against cybercrime". The report's diagnosis shows that the institutional framework for international collaboration on cybercrime is still at an early stage in Brazil. There are a few initiatives and actions. Notwithstanding this, a whole set of solutions need to be built up in the future. One of the crucial issues is the construction of a practical framework to receive collaboration from Internet service providers:

Informal cooperation exists with multinational ISPs [Internet Service Providers] voluntarily since they have no legal responsibility and are not obliged to answer requests from law enforcement unless they receive an official request (e.g. a court order or a search warrant). Currently, Brazil is developing a bilateral agreement between ISPs and law enforcement that allows ISPs to share data directly with the law-enforcement authorities. (…). At the operational level, exchanging information with foreign law-enforcement agencies and courts was effective. However, problems arise when requesting information from ISPs overseas and private Internet companies (such as Facebook and Google) in the US since they rarely respond and avoid cooperating with law enforcement in Brazil. In other words, if companies are based in Brazil, requesting the information is easier because they must comply with Brazilian law. Another concern raised during the review was an issue regarding Mutual Legal Assistance Treaties (MLATs) because they are very slow, which slows down investigations. It often takes as long as two years to get an answer to an official request from the US because they have only a few prosecutors dealing with MLATs from around the world (Global Security Capacity 2020).

Brazil ratified the Council of Europe's Convention on Cybercrime by promulgating a Congress Decree (Brazilian Federal Senate 2021). To accede the treaty, one more step is ongoing, the deposit. The country received an invitation to accede in 2019 (COU 21a). It is essential to describe the legal framework of the Convention a little to show that Brazil has characteristics that allow it to join the group of countries that are parties in the treaty. One tool of the Council of Europe about cybercrime is the Country Wiki, which is constantly evolving and describes the current position of many countries about cybercrime (COU 21b). The Convention on cybercrime has two main subjects. The first is a list of criminal conduct that the country must have in law. The second is a set of legal and institutional mechanisms for cooperation. Table 4.3 summarizes the Council of Europe's assessment of Brazil to show that its national criminal law proscribes most crimes (Council of Europe 2020).

Legal proscriptions on the Cybercrime Convention		Criminal proscription on the Brazilian Law: criminal code (CC), computer programs as intellectual property statute (CPIP), children and teenager protection statute (CTS), telephone and informatics communications' interception act (TICIA)	
Article 2	Illegal access	Articles 154-A and 154-B (CC)	It is an invasion of an information device (public or private)
Article 6	Misuse of device		
Article 3	Illegal interception	Article 10 (TICIA)	Interception without judicial authorization
Article 4	Data interference	–	No proscription
Article 5	System interference	Article 313-B (CC)	It is the unlawful modification or alteration of information systems
Article 7	Computer-related forgery	Article 297 (CC)	Forgery of public documents
		Article 298 (CC)	Forgery of private documents
		Article 298, sole paragraph (CC)	Forgery of credit or debit cards
		Article 313-A (CC)	Insertion of forged data in information systems
Article 8	Computer-related fraud	Article 171 (CC)	Larceny by fraud
		Article 155 (CC)	Steal by fraud
Article 9	Offenses related to child pornography	Article 240 (CTS)	It is the producing or reproducing of explicit content involving children or teenagers
		Article 241 (CTS)	Offer, trade, publish or distribute explicit content involving children or teenagers
		Article 241-A (CTS)	Offer, trade, publish or distribute explicit content involving children or teenagers using computers or networks
		Article 241-B (CTS)	Buy, have possession or store explicit content involving children or teenagers
		Article 241-C (CTS)	Simulate the participation of children or teenagers in explicit content
Article 10	It is an offense related to infringements of copyright and related rights	Article 184 (CC). Article 2 (CPIP)	It is a violation of copyright and related rights
Article 11	Attempt and aiding or abetting	Article 14 (CC)	The attempt to produce criminal conduct is punishable

Table 4.3. *Correlation of the Cybercrime Convention and Brazilian law. Source: Council of Europe Country Wiki, after authors' treatment*

So, as we explained, Brazil is joining the multilateral Convention on Cybercrime (Treaty 185). However, Brazil has not yet signed the additional protocol concerning the criminalization of racist and xenophobic acts committed through computer systems (Treaty 189) to date (March 2022) (COU 21c). Also, Brazil has still no accession to the Council of Europe's Convention on protecting children against exploitation and sexual abuse (Treaty 201). The Council of Europe's Convention on cybercrime is an essential tool to enable improvements in international cooperation regarding this subject. One of our informants, LEP-4, explains this. Despite being an LEP, she works on the administrative staff that carries out measures concerning international cooperation for investigations and prosecutions:

> In the Secretariat, I maintain contact with many colleagues because they always have doubts about whether they need some formal international cooperation. (…). I support and tell them when it is necessary to start an international cooperation procedure and how to do so. I can give an example. Brazil is pledging to accede to the Convention of Budapest. It is now in the National Congress for approval. The Budapest Convention keeps a network that functions non-stop to help international cooperation requirements. So, all member states must stay over nightly vigilantes. Without the Budapest Convention, Brazil has access to some other networks through the federal police department. Through those international police networks, it is possible to pledge some data. It is a hasty procedure. It must be this way because the evidence can disappear. We are talking about electronic data as evidence.

All the informants have the same point of view about the Convention on Cybercrime. They share the assessment that this is a significant path to developing a new national framework. However, sometime before the accession, Brazilian criminal law experienced an evolution to adapt itself to the digital age. These changes come from both some statutory reforms and judicial doctrines. Also, one can see some investment to strengthen the country's institutional capacities. It is no surprise that Brazilian law has a great deal of adherence to the international scenario when proscribing social conduct as cybercrime. The Council of Europe report also shows some evolution of the Brazilian law regarding communications interception (Telephone and informatics communications' interception act, or Federal statute No. 9,296 of 1996) and legal standards for storing and collecting digital data on Internet service providers (Internet rights act, or Federal Statute No. 12,965 of 2014).

Notwithstanding this, the most significant gap is still about legal rules for international cooperation. The Council of Europe's report mentions legal rules for cooperation in the Civil Procedures Code (Article 26 of the Federal Statute No. 13,105 of 2015). Indeed, that is not the best solution because criminal investigation and prosecution demand a specific legal basis (i.e. rules in the Criminal Procedures code). However, these changes in the Brazilian law also come from the dialogue between lawmakers and law enforcement staff, as pointed out by LEP-2:

> In the national group against cybercrime, we were only keeping track of most of the child pornography and racism crimes. The group also demands to keep track of crimes for illegal access to electronic devices. We have seen an apparent increase in this kind of crime because of the pandemic. In addition, Brazil will be at the final phase of accession to the Convention on Cybercrime. It is the sole international treaty about cybercrime. It proscribes a lot of specific cybercrimes that are still not in Brazilian federal statutory law. However, Brazil will have to pass those crimes into law shortly.

The recent accession is essential to the LEP. However, as we can see, Brazil will need to make many statutory reforms to adapt its national legal system. The Covid crisis gave a major push to Brazil's future accession. However, after this, one can hope that the next few years will be full of many federal statutory law initiatives about cybercrime. Adding international integration with the internal adaptation equals more investments to tackle cybercrime. The final accession will provide solutions to a wide array of problems, as LEP-2 describes:

> We believe that many more pledges for international cooperation will come from abroad after Brazil accedes to the Budapest Convention. (…). The accession will be essential to gather electronic evidence. After the accession, they will have the means to reach us. Moreover, we will have access to many countries that hold Internet infrastructure: European Union member states, the United States, Canada, Japan, and Australia. Those are the leading developed countries where most providers are. When

they pledge something from us, they need a contact point here. (...). Through those new contact points, they will be able to reach prosecutors and judges. They will be able to make more direct pledges to hasten the cooperation. Brazilian judges and prosecutors do not have this expertise. Those points will operate like liaison judges. Those judges always facilitate the relationships. The demand will rise both because procedures will be smoother and because of the increase of partner countries.

Notwithstanding this, these new possibilities of improving cooperation will demand more personnel and more training. Most police officers, prosecutors and judges have no special skills to afford the task of daily cooperation in criminal affairs. One possible solution is to create more cooperation points. Focus on training some LEP to assist others when dealing with international criminal cooperation. LEP-3 describes another vital aspect. The countries must shift their paradigm of international cooperation from bilateral treaties to multilateral ones, providing an effective creation of a cooperation network. Many cybercrimes usually happen in many jurisdictions at the same time. Having access to a wide range of member countries in the same treaty turns out to be a better solution than having many different treaties and, therefore, distinct procedures:

> The mistake is the bilateral point of view. Brazil should do the same as other countries by acceding to multilateral treaties. The best way to do this is to accede to other Council of Europe treaties, as we did with the Budapest Convention. (...). If the Brazilian government shares this perception, we must accede to many more. (...). The Budapest Convention is a good example. It has 65 members. So, after acceding to it, Brazil will amplify its capacities a lot. It is a tragic error to invest in bilateral treaties. It takes much energy to negotiate them, and only a tiny relationship comes from it. When a country accedes to a Council of Europe treaty or another regional treaty, which is open to third parties, its range of action expands a lot. Even the Interamerican treaties receive accession from non-American countries, like the Czech Republic and Israel. Many other countries are coming to accede to the Organization of American States treaties because they also are open to regional partners.

It is important to mention that another road to improve cooperation could be creating the BRICS framework for cooperating toward Internet policies, including cybersecurity. The governments of Brazil, Russia, India, China and South Africa debated this topic during the fifth BRICS summit in 2013. In the major statement, they stated:

> We recognize the critical positive role the Internet plays globally in promoting economic, social, and cultural development. We believe it's important to contribute to and participate in a peaceful, secure, and open cyberspace and we emphasize that security in the use of Information and Communication Technologies (ICTs) through universally accepted norms, standards, and practices is of paramount importance (BRICS Information Center 2013).

We can rely on the work of Luca Belli about the subject. He also uses the concept of legal interoperability in a similar way that we do here. For him, the BRICS are converting themselves into CyberBRICS. This digital turn has a clear relationship to a higher degree of policymaking toward cooperation. The first steps have their basis in science, technology and innovation efforts to cooperate, along with some economic integration. Notwithstanding, all those convergences can have a positive impact on cybersecurity. The author indicates five dimensions to cybersecurity policies – data protection, consumer protection, cybercrime, the preservation of public order, and cyber-defense. His point of view is optimistic: "Growing cooperation and legal interoperability amongst BRICS countries concerning with digital policy is increasingly possible and, to some extent, already happening" (Belli 2019). Nonetheless, Adriana Abdenur is more skeptical. She points out that the integration between the BRICS is still slow (Abdenur 2017). As a matter of fact, of the five BRICS countries, only South Africa is part of the Budapest Convention. Brazil is acceding to it right now (2021–2022). Despite being a state member of the Council of Europe (1996–2022) in the past, Russia never signed the treaty. The country withdrew from this international organization in March 2022. There is no information about efforts from India and China to accede to the Convention. At last, it is important to mention that new negotiations are starting within the United Nations to create another multilateral treaty for cooperation against cybercrime. These initiatives from BRICS and the UN are interesting, but only time will tell whether they will be fruitful.

After acceding to the Budapest Convention, Brazil will have much legislation to reformulate in the next few years and it will have to expand its budget on international cooperation to fight against cybercrime. The Covid crisis shows that issues like cybercrime and pandemics could not stay within borders. The accession to the Budapest Convention is direct proof that Brazil is learning this lesson. Nevertheless, this integration is a work in progress. The following section will describe how the Brazilian authorities have cooperated internationally without having comprehensive and fully effective statutory tools.

4.6. Managing international cooperation without having the best tools

The problem of not having a comprehensive framework of material legislation and procedural rules affects Brazil. Brazilian authorities have been adapting local norms to enable international cooperation. The role of direct cooperation has been a reliable way to do this. In this section, we will describe some cases and situations. After this, we will share concerns about the lack of efficiency without having comprehensive legal tools for international cooperation. LEP-1 describes the first case:

> If you allow, I will talk about a case that illustrates the issues of international cooperation. It is a case about a hacker forum. It was on some island, which is a cyber-paradise. It was a forum with at least 140 hackers from all over the world. One could only access the forum by the TOR system. The forum was closed. (…). The case comes from a very successful police infiltration. The FBI managed to put an agent inside the forum, and under a largely secure environment, these people who were so skilled lost their grip. Some did not mask their IP numbers, and the agent tracked them. A significant number of them. What did the United States do? They spontaneously shared the information with many countries, Brazil among them. There were two Brazilian hackers in the forum. This target bought and sold data from bank clients. (…). Thanks to the automatic transmission and local investigations, a simultaneous operation started in sixteen countries, and we arrested people. (…). The Brazilian target is in prison today. This hacker's imprisonment is very rare for property crimes. Unravelling the case, we found

> a sophisticated system to whitewash illegal profits. They were whitewashing with cryptocurrencies. We found pedophilia material, so he was also prosecuted for this. (…). We estimate that he had ten million US dollars in cryptocurrencies. (…). It would be impossible to reach him by traditional ways unless we did what the FBI did. That case is just an example and perhaps a powerful one. We had an increase in digital thefts during the pandemic. Brazil had a coronavirus aid for poor people. We saw a significant increase of frauds getting this money with false data, even in the name of dead persons.

The transcription is a bit long. However, the case is a paradigmatic one. It is essential to note that all the cooperation happens without straightforward procedural tools. In the future, this will have to change. It is crucial to have statutory provisions to grant full legal status to information from officials abroad. The spontaneous sharing of information about criminal activities between foreign investigators is not new. However, the central issue would be active cooperation between foreign authorities on ongoing investigations. The current tools available in international law are not enough to guarantee this. The Brazilian statutory law also does not have explicit provisions to grant this. The second case comes from the criticism of having sole central authorities process all the pledges of international cooperation, as LEP-2 explains. The following description strikes hard when someone thinks that even the regular crimes happen over the borders of big countries like Brazil:

> The first key point is the cooperation right on Brazil's border. It is the raw reality. Everyone needs to cooperate on the border. We have incredible stories of law enforcement personnel who share subpoenas over the radio not to cross the border and not to have to pledge cooperation through the central authority in Brasilia. Alternatively, even law enforcement personnel simply cross the border to deliver the subpoena (…). The same happens with judges. It is an illusion to think about central authorities and bureaucracy in a country with seventeen thousand kilometers of borders. Also, it has twenty-eight twin towns that are conurbations with foreign countries. (…). How could you think about asking about procedures in Brasilia, Lima, Bogotá, and even Paris? If we think about the frontier cooperation between Brazil and French Guiana (…). Sometimes having direct cooperation mechanisms is just necessary.

This explanation is quite clear. When LEP from different countries are close to the border, they may try to establish a way to cooperate that does not necessarily involve the bosses working in offices in their capitals. There is a way to solve this issue. Just before, one informant talked about the possibility of creating different points of contact to decentralize international cooperation to fight cybercrime. These new contact points will demand a more integrated national system and more trained LEP. Again, the rise of cybercrime that came along with the Covid crisis brought those issues to the table. This increase shows the importance of legal tools to cooperate quickly, as LEP-2 explains:

> The second key point is that most central authorities are not up to date to deal with cybercrime, which operates quickly. The bureaucracy happens to ignore situations that demand urgent measures. Cooperation by the traditional means is simply not adequate for some crimes. (…). Neither the Brazilian law nor any treaty we acceded to prescribes a straightforward tool for urgency and precautionary measures. We see some straightforward procedures like those from Operation Dark Code. The Federal Appellate Court had to decide whether the use of proof obtained with spontaneous cooperation was lawful. We recently received information from US agencies about frauds to allow illegal lumber trafficking from the Amazon. The prosecution filed a lawsuit in the Brazilian Supreme Court. The US Fish and Wildlife Service sent copies of the alleged fraudulent export licenses to the Brazilian Federal Police Department. The Brazilian Justice enacted warrants using the evidence from a direct police communication. I could go on telling you cases about this kind of direct police cooperation. We got the right design. (…). We cannot render evidence unlawful because we do not have channels to cooperate.

She talks about expanding legal channels through which evidence may potentially flow. The traditional model would centralize international cooperation in just one node to facilitate accountability. This model has some virtues. It focuses on having legal ways to prevent abuses of international cooperation. Although, it has some drawbacks. The most significant is that some cybercrime happens too quickly, and not being able to keep up with the pace may give the criminals the chance to escape. In addition, it does not help improve the country's capacity to thoroughly fight against cybercrime because just a few officials will have daily contact with the subject. LEP-3 also talks about the hardship of managing international cooperation even in cases that fall as a priority for them, like fighting against child abuse, child pornography and pedophilia:

> We work a lot against child pornography. When we ask for cooperation, the answers do come. However, the information always arrives late. We contacted the US office about cybercrime that work directly answering international cooperation pledges. They even created a unit to answer cybercrime cooperation pledges two years ago. However, it is just a unit located in Washington to respond to pledges worldwide. The big companies say they send information to this unit within a twenty-four-hour frame. (...). However, they take from six months to a year to respond. The answer is not necessarily positive. When the information arrives in Brazil, everything must pass through a translation into Portuguese. It takes some more time. The Brazilian unit within our Ministry of Justice states that the average time of the process is one year and eight months. Also, they indicate that the answer is positive only in two percent of the pledges. Our direct orientation to all colleagues is to avoid formal international cooperation measures. We suggest they contact the enterprises which have offices in Brazil directly. We need to rely on international cooperation only when the data is outside our border, and we cannot have any Brazilian legal tool to reach it (…).

It looks apparent that it would be better to have international cooperation capacities to fight cybercrime in Brazil and internationally. This explanation from LEP-3 tells us how international cooperation to fight cybercrime could be undermined by management constraints. These problems can happen in many situations. Lacking qualified staff to process all the pledges quickly is just one example like she said. In addition, more basic capacity is crucial. The digitization of some paperwork could be an issue, especially in some local jurisdictions that do not have proper resources. The LEP-2 tells a curious case about how the Covid lacked the digitization of some evidence for international cooperation:

> Right now, I am with a pledge for international cooperation from abroad. (...). The significant problem is that the judicial

process is not already digital to send a copy. Just to give an idea, this pledge came for us by the end of 2019. The Swiss authorities want information about all the prosecutions against a defendant. They had five urgent measures in the pledge. When I finally got the court's authorization to send a copy, the pandemic crisis struck. The court closed, and the material was physical. Two years later, the Swiss authorities asked again for the material. They say that the proofs will still be helpful. Then I contacted the judge directly, and they will digitize everything. I will answer them in a few days. The Secretariat for International Cooperation is demanding efforts for all federal courts to digitize all the processes. One of the reasons is to facilitate international cooperation.

Notwithstanding this, the problems to enable international cooperation about cybercrime are not only about management, as one can see. We have a legal problem that also needs a solution, not only in procedural law. Striking results will be tough to obtain in a short timeframe. There is a necessary legal and social debate about how felonies and misdemeanors, as legal concepts, may have a different interpretation, depending on each national legal system. However, it is imperative to establish a basis to debate this, as the next section shows.

4.7. Difficulties with cooperation: joints, mortises, and notches

Section 4.3 showed a theoretical model to understand that international cooperation is not a secondary matter to cybercrime. It is a vital part of the solution. We also explained that we have two axes to measure the degree of interoperability when dealing with international cooperation. In the two previous sections, the LEP talked a lot about procedures. However, LEP-3 shared one crucial piece of information. She said that the Brazilian Ministry of Justice states that they have success in only two percent of the pledges for cooperation. There are many possibilities to explain this low score of success. One explanation relates to material law, that is, how the countries proscribe some social conducts as crimes.

To through some light on this issue, we will describe the limits in legal systems' interoperability with one well-known case: LICRA versus Yahoo (Greenberg 2003). This case started when two French civil society associations filed lawsuits against Yahoo in France. This search engine was advertising Nazi memorabilia in its auction marketplace. However, similar to many countries, in France this conduct is a crime. The League against racism and antisemitism (LICRA, acronym in French) and the Jewish student's union of France (UEJF, the same) demanded that Yahoo should block this content, at least in France. In its defense, Yahoo replied that its services' provision was international. Therefore, it could not block such information, considering just one country's law.

Moreover, Yahoo defended itself by stating that the civil society associations had no power to file a lawsuit in France against an American Internet service provider. Also, it argues that the First Amendment of the United States constitution would block any judicial ruling against Yahoo, relying on the Freedom of Speech rights and State Action Doctrine. The French court ruled that Yahoo was liable for allowing a crime to happen on French soil. Yahoo did not file any appeal against the French ruling. Then, a year after, Yahoo filed a lawsuit against LICRA and UEJF in a federal district court in California. The district court granted the pledge at first. Its ruling stated that the French decision could not have legal validity in the United States. Both LICRA and UEJF filed appeals against such understanding. The circuit opinion reviewed the first ruling to state that the federal judicial branch could not exercise jurisdiction against LICRA and UEJF. Therefore, the French ruling had validity in France. Some years later, the case reopened, and the same court again examined the Yahoo allegations. The conclusion was broader than the first. It ruled that one cannot extraterritorially enforce the First Amendment of the United States Constitution. This classic example brings two different sets of legal issues. The first set comprises the legal definition of conduct as a crime. Even when the definitions seem similar, some significant differences may apply. The second set is the vast array of difficulties that overseas lawsuits pose to the national legal systems. A hypothetical crime may happen in France. However, in the United States, the same conduct may not be legally a crime. How can one enforce a French ruling against an American defendant? One interesting example comes from the United States federal statutory law. The Title 18 of the US Code regulates criminal law and procedures. One case of extraterritorial crime that an American citizen may perpetrate abroad and suffer punishment from the US federal jurisdiction: abuses against minors (United States Department of Justice 2020). This situation is exceptional to the usual trend that classifies the federal crimes of Title 18 as traditionally territorial.

This legal problem hugely affects international cooperation. Considering its national law, how can an authority share data about a case that does not deal with a crime? This situation affects crimes of racism and even situations like Nazi propaganda, as the LICRA versus Yahoo case shows. LEP-3 informs us that Brazil is giving up on trying to use traditional international cooperation tools for some situations, which the United States does frame as crimes:

> We avoid using international cooperation because of the long time to receive answers and the lack of response. One example comes from defamation. The other comes from racism. The leading Internet content providers are in the United States. However, the US authorities will not answer pledges for cooperation because those conducts are not crimes in their law. One requirement is that the conduct must be criminal both here and there. Defamation is not a crime in the United States. Racism falls under a particular crime, requiring many legal details, like the criminal's malicious intent. So, the primary orientation of our unit to those cybercrimes is to give up on trying international cooperation.

For one side, it is shocking to see the idea of giving up requiring evidence to investigate a crime mainly because the United States does not frame it as so. That is the central theoretical problem that the legal community will have to address in the following years, along with the refinement of the tools for direct international cooperation in investigations. LEP-2 describes that the direct assistance of the Big Tech is helping:

> Over the years, we have built a relationship with those Internet service providers to facilitate investigations. Nowadays, we have a close relationship with most of their leading representatives here. With WhatsApp, we pose direct and specific questions. I had a doubt two weeks ago. I sent a direct message, and they collaborated. The last service to enable this collaboration was WhatsApp. Facebook currently represents it. (...). There is a list of data that they can share with the authorities. There is no problem with receiving metadata. This metadata is not encrypted because, in the United States, a warrant is not necessary to give away an IP number, for example. So, they have no problem sharing this. No resistance.

They try to collaborate in the best way possible and specifically about some crimes that also affect the United States.

On one side, this direct way of cooperating with Big Tech allows the LEP to further advance some investigations without experiencing the drawbacks that the international cooperation systems pose. Nonetheless, on the other side, it is not a complete solution for two reasons. The first is that the foreign enterprises maintain their current compliance status only with the United States law. The second one is that this situation does not help to improve interoperability between systems. Furthermore, it strengthens the idea that the United States material law must be universally applied abroad. However, there are even procedural problems, as LEP-2 informs us:

> The bigger problem with some content is a rigorous judicial doctrine about warrants in the United States. They have special requirements, like the probable cause doctrine and so on. Therefore, they disagree that some Brazilian judicial warrants follow the view of a typical American judge. The providers also explain that they are quite afraid of receiving punishment for giving away some data unlawfully. However, this has never happened. A paper about the CLOUD Act explains what we call service effects. The United States' service providers must comply with the American legal system. However, the Brazilian federal law also determines that they must comply with our legal system.

There are no clear skies for international cooperation to fight against cybercrime. The United States' procedural requirements will appear even when Brazilian authorities ask for direct collaboration from Big Tech. One can have a degree of criticism about the collaboration of service providers directly with law enforcement personnel. Nonetheless, the actual legal issue is also a theoretical problem. Creating a higher interpretative solution level is necessary to integrate investigation and prosecution activities legally and effectively tackle cybercrime. In conclusion, we will explain a bit about the prospects.

4.8. Conclusion: what to expect from the future?

Relying on the interviews and interpreting them, we can have two significant prospects. The first prospect is internal to the Brazilian legal system. The rise of cybercrime during the Covid crisis shows a clear path for Brazil: it must deepen its international integration and create national statutory provisions to integrate itself better. There is some homework to do, to put it simply. As LEP-2 says: "Our legal framework is not complete (...). We do not have a specific statute to criminal cooperation". She also says that Brazil must pursue the integration between all countries in the Americas by effectively establishing Ameripol (Ameripol 2022). This organization is still institutionalizing: "Currently, Ameripol does not have a proper structure. It is a basis for a future international organization. It has a treaty. Brazil is lagging, and other countries are too. It already exists in Bogota".

The second prospect is international. Brazil and the other countries must endorse the construction of an international regime to empower active, quick and effective cooperation between investigators, prosecutors and judges. Along with the international regime, there are some regional efforts, some initiatives like the BRICS cooperation, and even a debate about a new United Nations treaty starting within this global international organization (Rodriguez and Baghdasaryan 2022). On a regional level, the main idea of LEP-2 is projecting a system like Europol: "It must happen like the European Union does, despite not being a federation. This is how it works between Portugal, Italy, Germany and France – direct cooperation without intermediaries. However, how could this happen worldwide?" She also has an answer to that:

> We must find at the international level a solution for lawfully exchanging proofs. That solution must come from a system of direct cooperation. We have human rights treaties that oblige this. The states must comply with their human rights obligations in investigating, prosecuting, and punishing cybercrime. If those obligations are mandatory, they all demand international cooperation. Furthering this, we may see an improvement in judicial and police diplomacy.

All these international, regional and even innovative agendas, like the BRICS, point in the same direction. But nothing is ever so easy to the archive when looking forward to international cooperation. Nevertheless, all those achievements require a high degree of interoperability between legal systems. This debate about legal interoperability will demand much effort to attract legal experts to find solutions to refine various legal systems. This solution will not come without struggle, as LEP-2 says: "Then we will see the resistance of the transnational enterprises under one or other legal system". This solution is nothing new. All changes must come from a lot of hard work and qualified debate. The Covid crisis is showing everyone that global threats require international convergent solutions. To tackle cybercrime, we will also require an international interoperable regime.

4.9. References

Abdenur, A. (2017). Can the BRICS cooperate in international security? *International Organizations Research Journal*, 12(3), 73–95.

Adams, J.-A.M. (1996). Controlling cyberspace: Applying the computer fraud and abuse act to the internet. *Santa Clara Computer & High Tech. Law Journal*, 12(2), 403–434.

Ameripol (2022). Comunidade de Policías de América – integración para la protección y seguridad ciudadanos [Police community of Americas – integration for protecting and securing citizenship] [Online]. Available at: http://www.ameripol.org [Accessed 1 July 2022].

Belli, L. (2019). CyberBRICS: A multidimensional approach to cybersecurity for the BRICS. In *CyberBRICS: Mapping Cybersecurity Frameworks in the BRICS*, Belli, L. (ed.). Fundação Getúlio Vargas, Rio de Janeiro [Online]. Available at: https://cyberbrics.info/wpcontent/uploads/2020/01/DIGITAL_CyberBRICS_Pre_launch_Brochure.pdf [Accessed 1 July 2022].

Belli, L. and Foditsch, N. (2016). Network neutrality: An empirical approach to legal interoperability. In *Net Neutrality Compendium: Human Rights, Free Competition and the Future of the Internet*, Belli, L. and De Filippi, P. (eds). Springer, Cham.

Benbouzid, B. and Ventre, D. (2016). Pour une sociologie du crime en ligne. *Reseaux*, 197–198, 9–30.

Benson, C., Jablon, A.V., Kaplan, P., Rosenthal, M.H. (1997). Computer crimes. *American Criminal Law Review*, 34(2), 409–444.

Brazilian Federal Senate (2021). Aprova o texto da Convenção sobre o Crime Cibernético, celebrada em Budapeste, em 23 de novembro de 2001 [Approves the text of the Convention Against Cybercrime signed on Budapest on 23 November 2001], Brasília [Online]. Available at: https://legis.senado.leg.br/norma/35289207 [Accessed 1 July 2022].

BRICS Information Center (2013). BRICS and Africa: Partnership for development, integration, and industrialization – eThekwini Declaration – 27 March 2013, University of Toronto [Online]. Available at: http://www.brics.utoronto.ca/docs/130327-statement.html [Accessed 1 July 2022].

Christensen, K.D. (1997). Fighting software piracy in cyberspace: Legal and technological solutions. *Law and Policy in International Business*, 28(2), 435–476.

Council of Europe (2020). Brazil: Cybercrime legislation – Domestic equivalent to the provisions of the Budapest Convention. Strasbourg [Online]. Available at: https://rm.coe.int/octocom-legal-profile-brazil-final/1680a11cb5 [Accessed 25 May 2020].

Council of Europe (2021a). Congress of Brazil approves accession to the Budapest Convention. Strasbourg [Online]. Available at: https://www.coe.int/en/web/cybercrime/-/congress-of-brazil-approves-accession-to-the-budapest-convention [Accessed 16 December 2021].

Council of Europe (2021b). Country Wiki. Strasbourg [Online]. Available at: https://www.coe.int/en/web/octopus/country-wiki [Accessed 1 July 2022].

Council of Europe (2021c). Complete list of the Council of Europe's Treaties. Strasbourg. [Online]. Available at: https://www.coe.int/en/web/conventions/full-list [Accessed 1 July 2022].

European Union (2017). New European interoperability framework: Promising seamless services and data flows for European Public Administration. Brussels [Online]. Available at: https://ec.europa.eu/isa2/sites/default/files/eif_brochure_final.pdf [Accessed 1 July 2022].

Federal General Prosecutor's Office of Brazil (2022). Crime de invasão de dispositivo informático [Electronic intrusion investigations and prosecutions], crime previsto no Estatuto da Criança e do Adolescente [crimes against children investigations and prosecutions], crimes resultantes de preconceito, raça ou cor [racial crime investigations and prosecutions] [Online] Available at: http://www.mpf.mp.br/atuacao-tematica/ccr2/sobre/boas-praticas/dados_preconceito-eca-sigilo.pdf [Accessed 4 April 2022].

Global Cyber Security Capacity Center (2020). Cybersecurity capacity review: Federative Republic of Brazil. Organization of American States (OAS), Washington, DC [Online]. Available at: https://cybilportal.org/projects/cmm-review-brazil-as-part-of-the-digital-access-programme-trust-resilience [Accessed 1 July 2022].

GovInfo (1984). Counterfeit Access Device and Computer Fraud and Abuse Act, Public Law. 98th Congress, 12 October 1984 [Online]. Available at: https://www.govinfo.gov/content/pkg/STATUTE-98/pdf/STATUTE-98-Pg1837.pdf [Accessed 1 July 2022].

Grabosky, P. (2003). Cibercrime. *Cadernos Adenauer*, Konrad-Adenauer Stiftung, 4(6), 47–79 [translation to Portuguese: Alexandre Veronese].

Greenberg, M.H. (2003). A return to Lilliput: The LICRA v. Yahoo! Case and the regulation of online content in the world market. *Berkeley Technology Law Journal*, 18, 1191–1258.

Impreza (2021). The words that led the president of SaferNet to be spied on, threatened, and to leave the country. *Impreza* [Online]. Available at: https://impreza.host/the-words-that-led-the-president-of-safernet-to-be-spied-on-threatened-and-to-leave-the-country [Accessed 11 December 2021].

Internet Archives (2022). WayBackMachine [Online]. Available at: https://web.archive.org/web/20190618070639/http://indicadores.safernet.org.br/index.html [Accessed 18 June 2018].

Knoth, P. (2022). Denúncias de neonazismo na Internet aumentam 60% e batem recorde [Reports against Neo Nazism on the Internet rise 60%, and breaks record] [Online]. Available at: https://tecnoblog.net/noticias/2022/02/09/denuncias-de-neonazismo-na-internet-aumentam-60-no-brasil-e-batem-recorde [Accessed 9 February 2022].

Küster, M.W. (2012). Standards for achieving interoperability of e-Government in Europe. In *Digital Democracy: Concepts, Methodologies, Tools, and Applications*, Information Resources Management Association (ed.). IGI Global, Hershey, PA.

Levy, S. (1984). *Hackers: Heroes of the Computer Revolution*. Anchor Press/Doubleday, New York.

Lima, R. (2020). Government recovers BRL 42 million from coronavirus aid fraud. Brazilian Report [Online] Available at: https://brazilian.report/liveblog/coronavirus/2020/06/26/coronavirus-aid-fraud [Accessed 28 June 2020].

Loiseau, H. (2020). The "science" of cybersecurity in the human and social sciences: Issues and reflections. In *Cybersecurity in Humanities and Social Sciences: A Research Methods Approach*, Loiseau, H., Ventre, D., Aden, H. (eds). ISTE Ltd, London, and John Wiley & Sons, New York.

Mellone, M. (2014). Legal interoperability in Europe: An assessment of the European payment order and the European small claims procedure. In *The Circulation of Agency in e-Justice*, Contini, F. and Lanzara, G.F. (eds). Springer, Cham.

Mitchell, S.D. and Banker, E.A. (1998). Private intrusion response. *Harvard Journal of Law & Technology*, 11(3), 699–732.

Morando, F. (2013). Legal interoperability: Making open government data compatible with businesses and communities. *Italian Journal of Library, Archives and Information Science*, 4(4), 441–452.

Muggah, R. and Thompson, N.B. (2015). Brazil's cybercrime problem: Time to get tough [Online]. Available at: https://www.foreignaffairs.com/articles/south-america/2015-09-17/brazils-cybercrime-problem [Accessed 17 September 2020].

Oliveira, T.T.N. and Almeida, C.S.B. (2014). How offenders use the web in Brazil to spread sexual abuse material: Findings from ongoing evidence-based research. UNICEF Office of Research – Innocenti [Online]. Available at: https://www.unicef-irc.org/article/841-how-offenders-use-the-web-in-brazil-to-spread-child-sexual-abuse-material-findings.html [Accessed 17 September 2020].

Rodrigues, D. (2020). Governo recupera só 0,4% do valor estimado em fraudes no auxílio emergencial [Brazilian government only retrieves 0.4% of estimated coronavirus emergency financial assistance frauds]. *Poder 360* [Online]. Available at: https://www.poder360.com.br/governo/fraudes-no-auxilio-r-2274-milhoes-foram-devolvidos-aos-cofres-publicos [Accessed 14 December 2020].

Rodriguez, K. and Baghdasaryan, M. (2022). UN Committee to begin negotiating new cybercrime treaty amid disagreement among states over its scope. Electronic Frontier Foundation [Online]. Available at: https://www.eff.org/deeplinks/2022/02/un-committee-begin-negotiating-new-cybercrime-treaty-amid-disagreement-among [Accessed 15 February 2022].

Rolfini, F. (2020). Cibercrime: ataques no Brasil aumentam mais de 300% com a pandemia [Cybercrime: Attacks rise more than 300% along with the pandemic]. *Olhar Digital* [Online]. Available at: https://olhardigital.com.br/2020/07/03/seguranca/cibercrime-ataques-no-brasil-aumentam-mais-de-300-com-a-pandemia [Accessed 3 July 2020].

Rubin, H., Fraser, L., Smith, M. (1995). US and international law aspects of the Internet: Fitting square pegs into round holes. *International Journal of Law and Information Technology*, 3(2), 117–173.

SaferNet (2021a). Estatuto [By-laws] [Online]. Available at: https://new.safernet.org.br/content/estatuto [Accessed 16 December 2021].

SaferNet (2021b). Parceiros [Partnerships] [Online]. Available at: https://new.safernet.org.br/content/parceiros [Accessed 16 December 2021].

SaferNet (2021c). Indicadores [Indicators] [Online]. Available at: https://indicadores.safernet.org.br/indicadores.html [Accessed 16 December 2021].

SaferNet (2022). Indicadores [Indicators] [Online]. Available at: https://indicadores.safernet.org.br/indicadores.html [Accessed 1 July 2022].

Santosuosso, A. and Malerba, A. (2014). Legal interoperability as a comprehensive concept in transnational law. *Law, Innovation and Technology*, 6(1), 51–73.

Sterling, B. (1992). *The Hacker Crackdown: Law and Disorder on the Electronic Frontier*. Bantam Books, New York.

Swedberg, R. (1998). The economy and law. *Max Weber and the Idea of Economic Sociology*, Princeton University Press, Princeton, NJ.

Trindade, L.V.P. (2018). Brazil's supposed "racial democracy" has a dire problem with online racism. *The Conversation* [Online]. Available at: https://theconversation.com/brazils-supposed-racial-democracy-has-a-dire-problem-with-online-racism-99343 [Accessed 7 August 2018].

Trubek, D.M. (1972). Max Weber on law and the rise of capitalism. *Wisconsin Law Review*, 3, 720–753.

United Nations (1965). International Convention on the Elimination of all Forms of Racial Discrimination. UN Office of the High Commissioner of Human Rights, New York. 21 December 1965 [Online]. Available at: https://www.ohchr.org/en/instruments-mechanisms/instruments/international-convention-elimination-all-forms-racial [Accessed 1 July 2022].

United Nations (2000a). Optional Protocol to the Convention on the Rights of the Child on the sale of children, child prostitution, and child pornography. UN Office of the High Commissioner of Human Rights, New York, 25 May 2000 [Online]. Available at: https://www.ohchr.org/en/instruments-mechanisms/instruments/optional-protocol-convention-rights-child-sale-children-child [Accessed 1 July 2022].

United Nations (2000b). Protocol to Prevent, Suppress and Punish Trafficking in Persons Especially Women and Children, supplementing the United Nations Convention against Transnational Organized. UN Office of the High Commissioner of Human Rights, New York, 15 November 2000b [Online]. Available at: https://www.ohchr.org/en/instruments-mechanisms/instruments/protocol-prevent-suppress-and-punish-trafficking-persons [Accessed 1 July 2022].

United States Department of Justice (2020). Citizen's guide to US federal law on the extraterritorial sexual exploitation of children. US Department of Justice, Washington, DC [Online]. Available at: https://www.justice.gov/criminal-ceos/citizens-guide-us-federal-law-extraterritorial-sexual-exploitation-children [Accessed 28 May 2020].

Veronese, A. (2021). Personal data and transborder flows between the EU and US: Dilemmas and potential for convergence. In *Extraterritoriality of EU Economic Law: The Application of EU Economic Law Outside the Territory of the EU*, Cunha Rodrigues, N. (ed.). Springer, Cham.

Wall, D. (1998). Catching cybercriminals. *International Review of Law, Computers and Technology*, 12(2), 201–218.

Wall, D. (1999). Cybercrime: New wines, no bottles? In *Invisible Crimes: Their Victims and their Regulation*, Davies, P., Francis P., Jupp, V. (eds). MacMillan Press, London.

Wall, D. (2007). *Cybercrime: The Transformation of Crime in the Information Age*. Polity Press, London.

Weber, R.H. (2014). Legal interoperability as a tool for combatting fragmentation. Chatham House [Online]. Available at: https://www.cigionline.org/sites/default/files/gcig_paper_no4.pdf [Accessed 1 July 2022].

Zeviar-Geese, G. (1998). The state of the law on cyber jurisdiction and cybercrime on the Internet. *Gonzaga Journal of International Law*, 1(1), 119–146.

4.10. Appendix: List of interviews and questions

We performed four interviews with federal law enforcement professionals (LEP). All of them have vast experience in the subject. All interviews were partially open to adaptation during the talks. Their names are not shown in this chapter to protect them. We also used a neutral or female gender for everyone to protect them. We codified the interviews by date as follows: LEP-1 (21 July 2021), LEP-2 (31 August 2021), LEP-3 (1 September 2021), LEP-4 (3 September 2021).

1) What is your position, and for how long have you been working in this position?

2) Please describe your job activities, and how they relate to cybercrime. Could you tell us in what degree do you use international cooperation in your daily activities?

3) In your current position, do you have any activities like investigations or criminal prosecution of cybercrimes on the Internet? What are the most recurring cybercrimes? Which ones do you consider the most important? Please explain.

4) Do you perceive any change during this current crisis period (Covid-19)? How do you compare the current cybercrime activities with the former time?

5) Considering your current position, what can you tell us about your actions investigating or prosecuting cybercrime, especially when you need to access digital evidence through cooperation mechanisms?

6) From your point of view, what are the significant issues to provide cooperation effectiveness to Brazilian investigations?

7) What kind of improvements concerning criminal investigations and digital evidence issues could Brazilian international cooperation experience?

5

Has Covid-19 Changed Fear and Victimization of Online Identity Theft in Portugal?

5.1. Introduction

The Internet plays a crucial role in daily life, allowing the use of several governmental, commercial, economic, financial, leisure and educational services (Grabosky and Smith 2001; Dias 2012). Therefore, nowadays, the Internet represents a pillar of the global economy, especially during and after the Covid-19 pandemic. Despite the undoubtable advantages that the Internet has brought, a set of criminal opportunities must be taken into consideration. In fact, with its fast and unprecedent development, new crimes have emerged, such as hacking, and others (traditional crimes) have been extended to cyberspace, such as different types of fraud (e.g. online identity theft (OIT)), cyberstalking or cyberbullying. This chapter will focus on the OIT, which is now considered one of the fastest growing online crimes (Golladay and Holtfreter 2016; Williams 2016), resulting in relevant financial losses for victims. Identity theft is the "unauthorized use or attempted use of an existing account (such as credit/debit cards, savings, telephone and online), the unauthorized use or attempted use of personal information to open a new account or misuse of personal information for a fraudulent purpose such as providing false information to law enforcement" (Harrell 2015, p. 2).

Chapter written by Inês GUEDES, Joana MARTINS, Samuel MOREIRA and Carla CARDOSO.

As previously mentioned, during the Covid-19 pandemic, the population changed their routines and multiple actions are now performed in cyberspace, leading to an increasing exposure to risks (Horgan et al. 2020). Consequently, it is important to understand that victimization rates have increased (Hawdon et al. 2020), and there is also a heightened worry and perceived risk of OIT due to the Covid-19 pandemic. Among the criminological theories that have been used to explain online victimization, routine activities theory (RAT) is one of the most empirically tested (Cohen and Felson 1979; Bossler and Holt 2009; Reyns 2013; Reyns 2015; Reyns and Henson 2015). Therefore, the present study, undertaken in the Portuguese context (2017 and 2021), intends to (i) analyze the levels of victimization, fear and risk perception of OIT before and after the Covid-19 pandemic crisis; (ii) analyze the evolution of online routine activities (exposition, target suitability and guardianship behaviors) also before and after the Covid-19 pandemic crisis; (iii) understand the victimization of other forms of online victimization during the last 24 months. The chapter will start with an overview of the impact of the Covid-19 pandemic on cybercrime, followed by the definition of OIT and the determinants of its victimization. Therefore, we will focus not only on the feasibility of the RAT to explain the victimization of OIT, but also on the importance of other individual variables. Finally, in the theoretical section we will describe the main variables that have been used to address fear and perceived risk of OIT. Next, the methods section will describe the survey and procedures used in the present study, followed by the obtained results and their discussion.

5.2. The impact of the Covid-19 pandemic on cybercrime

The Covid-19 pandemic and the associated lockdown measures imposed by governments to control the spread of the virus had a large impact on several social spheres, affecting people's routine activities worldwide. For instance, many businesses and education institutions moved their activities toward home, several physical retail stores closed, there was reduction in mobility and households were occupied most of the time. Consequently, many began using the Internet daily for work, education, leisure, socialization and consumption (Buil-Gil et al. 2021; Kemp et al. 2021; Buil-Gil and Zeng 2022). Based on the RAT (Cohen and Felson 1979), it has been claimed that these tremendous changes in routine activities brought on by the Covid-19 have influenced crime opportunities, displacing them from the physical to the cyberspace. More specifically, the shift in social activities from the physical to the virtual world increased the likelihood of converging motivated offenders and suitable targets (in the absence of capable guardians) in cyberspace (Hawdon et al. 2020; Payne 2020; Kemp et al. 2021).

In fact, on the one hand, studies (e.g. Mohler et al. 2020; Abrams 2021; Buil-Gil et al. 2021; Langton et al. 2021) have found evidence that many street crimes and crimes in outdoor physical spaces (e.g. drug crimes, vandalism, robbery and theft) decreased during the Covid-19 pandemic. Considering that people started spending less time on the streets and other public spaces, the researchers argued that opportunities for physical convergence in time and space of motivated offenders and suitable targets in the absence of capable guardians were reduced (Mohler et al. 2020; Buil-Gil et al. 2021; Kemp et al. 2021; Langton et al. 2021). On the other hand, since people started spending more time on the Internet, expanding the frequency, intensity and variety of activities online (e.g. working, studying, shopping and socializing), opportunities for cybercrimes increased (Mohler et al. 2020; Buil-Gil et al. 2021; Kemp et al. 2021). As a matter of fact, academic research has been demonstrating that cybercrime increased during the pandemic. To explore the effect of the pandemic on cybercrime in the United Kingdom, Buil-Gil et al. (2021) conducted a time series analyses of online crime data recorded by Action Fraud between May 2019 and May 2020. The results indicated that reports of cybercrime have increased during the Covid-19 outbreak, especially the amount of fraud associated with online shopping and auctions, hacking of social media and e-mail. Similarly, Kemp et al. (2021) applied a time series analysis to data on cybercrime reported to Action Fraud in the United Kingdom to examine a potential increase in online crime brought about by the Covid-19 pandemic. The results revealed that cybercrime increased after the introduction of lockdown measures beyond predicted levels by the time-series technique, but the changes in victimization were not homogeneous across fraud types. One the one hand, they noted a statistically significant increase in online shopping fraud, romance fraud and cyber-dependent crimes (e.g. denial-of-service attacks, hacking of social media and e-mail). On the other hand, they observed a significant decrease in ticket fraud, arguing that less ticket-related offline leisure and transport activities led to a decrease in online ticket fraud. The results of other studies are in line with the above-mentioned findings. In the United States, an increase in reports of most types of online fraud was noticed (e.g. imposter businesses, fraudulent text messages, online shopping fraud and romance fraud) in the first three months of 2020 compared to the same period in 2019 (Payne 2020). Consistently, Lallie et al. (2021) observed that, globally, cyberattacks became much more prevalent during the Covid-19 pandemic. However, Hawdon et al. (2020) conducted a study in the United States, comparing pre-pandemic rates with post-pandemic rates of cybervictimization, and found that, on the contrary, the cybervictimization rates did not change.

Although several studies have explored the effects of the Covid-19 pandemic on different types of cybercrime and found evidence that cybercrime generally increased, there is a lack of research on its influence, specifically on OIT. As noted by Kemp et al. (2021), cybercrime is an umbrella concept, encompassing a very diverse range of online crimes. Thus, the pandemic may affect victimization differently depending on the type of cybercrime. In that sense, scholars (e.g. Hawdon et al. 2020; Kemp et al. 2021) have highlighted the importance of a crime-specific approach to cybercrime research. In fact, studies (e.g. Buil-Gil et al. 2021; Kemp et al. 2021) have found that the pandemic has not had identical effects on different cybercrimes.

Hawdon et al. (2020) included OIT in their study, but did not find a significant difference in OIT rates pre- and post-pandemic. However, recent research suggests that the pandemic may have also led to an increase in this cybercrime. Buil-Gil and Zeng (2022) analyzed the extent to which online romance fraud known to the police increased during Covid-19 in the United Kingdom and which population groups were the most affected by this type of fraud. The results showed a large increase in online romance fraud and a higher uplift among young adults than older persons. Moreover, the researchers emphasized that in the perpetration process, fraudsters adopt fake identities of "ideal" partners, simultaneously exploring the increase in Internet use and the psychological adverse effects of the pandemic, such as loneliness. Indeed, cybercriminals often adopt a false identity to manipulate people into providing information or performing activities, mimicking, for instance, persons or entities that have authority, competence and integrity, or impersonating friends on social media to gain the trust of the victims. In that sense, identity theft is common when perpetrating, for example, several types of fraud or phishing attacks (Abbasi et al. 2010; Westerman et al. 2014; Algarni et al. 2017; Buil-Gil and Zeng 2022), that academic research has shown to have increased during the Covid-19 pandemic.

5.3. Evolution of cybercrime in Portugal

In Portugal, under the generic term of cybercrime, a heterogeneous set of legal types of crime is included. These are not only the crimes foreseen in the Portuguese Cybercrime Law (Law number 109/2009[1]), but also others described in the Penal Code and in separate legal sources. Besides these multiple legal categories where cybercrimes can be found, some offenses such as the case of OIT do not correspond to a specific type of crime, but a widen phenomenon to which different types of crime apply. Therefore, as the Portuguese Cybercrime Office argues[2], the statistical quantification of cybercrime is not easy (2020) since judicial statistics aggregate crimes according to their legal types (for instance, scams and crimes against copyright), not considering autonomously or separately those that occur online. A set of mechanisms have been used in Portugal to measure cybercrimes, such as the existence of an electronic address (since 2016) where individuals can send complaints related to illicit activities that have occurred online. These complains are then forwarded to the competent authorities in Portugal. Therefore, the Portuguese Cybercrime Office has been producing reports based on complaints received since 2016. Concerning the evolution of cybercrime in the Covid-19 pandemic, the reports launched by this governmental institution showed a progressive and persistent increase in complaints received. For instance, considering only the period between January and May 2020 (corresponding to the peak of lockdown), the complaints were much higher (139% of increasing) when compared to the total numbers of 2019. Moreover, they also observed that April 2020 was the peak of received complaints, which started to decrease again in May. There were four main online offenses reported by the Portuguese Cybercrime Office: fraud with MBWAY payments[3], fraud with email and SMS containing

---

1. Transposing the Framework Decision No. 2005/222/JHA, of the Council of the EU, of 24 February, on attacks against information systems into the domestic legal order. In addition, it adapts the Council of Europe Convention on Cybercrime, approved in Budapest on 23 November 2001.
2. Report available at: https://cibercrime.ministeriopublico.pt/sites/default/files/documentos/pdf/denuncias-de-cibercrime-25-01-2022.pdf [Accessed March 2022].
3. MB WAY is a payment solution that allows individuals to make purchases online and in physical stores, by generating MB NET virtual cards. The MB WAY app was developed in Portugal.

malware, phishing campaigns and email extorsion[4]. When analyzing the results related to the year of 2021, it is possible to observe that the complaints were higher in the periods of lockdown. However, the reports also conclude that regardless of the peaks of cybercrime during the lockdowns (both in 2020 and 2021), there has been a consistent and progressive increase in cybercrime in the last few years. In fact, while from the year 2019 to 2020 the increase was 288%, from 2020 to 2021 the increase in received complaints was 213%. In addition, it is important to note that the main online offenses reported in 2021 were phishing attacks, online fraud (e.g. online consumer fraud) and fraud with crypto assets and other financial products[5].

Another important source of information is the Risks and Conflicts report launched in 2021[6]. This report concluded that phishing/smishing attacks were the most relevant cyberthreats during the period of 2020 (more than 160% compared to 2019), followed by malware infections (increase of 37% compared to 2019). Moreover, offenses such as fraud using MBWAY and others that are related to human factor were the most prevalent. Finally, reports undertaken by the "Line Secure Internet"[7] launched by the Portuguese Association for Victim Support (APAV) also showed a growth in the number of calls received by that entity. While the peaks of reports were made in the periods of lockdown, a general increase in received calls can be observed (from 827 in 2019 to 1164 in 2020). Both in 2019 and 2020, phenomena such as fraud, identity theft, phishing and sextortion were among the top 10 of the most common offenses reported by victims.

In summary, the above-mentioned reports showed an increase in overall cybercrime during the Covid-19 pandemic. However, less is known about specific tendencies of OIT, both concerning victimization levels and fear/risk perception levels.

---

4. Report available at: https://cibercrime.ministeriopublico.pt/sites/default/files/documentos/pdf/2020_06_01_cibercrime_em_tempo_de_pandemia.pdf [Accessed March 2022].

5. Report available at: https://cibercrime.ministeriopublico.pt/sites/default/files/documentos/pdf/denuncias-de-cibercrime-25-01-2022.pdf [Accessed March 2022].

6. Report available at: https://www.cncs.gov.pt/content/files/relatorio_riscos.conflitos2021__observatorlociberseguranca_cncs.pdf [Accessed March 2022].

7. Report available at: https://apav.pt/apav_v3/images/pdf/Estatisticas_APAV_LinhaInternet Segura_2021.pdf [Accessed March 2022].

5.4. Online identity theft (OIT)

5.4.1. *Definition and modus operandi*

Cyberidentity can be defined as "the set of physical, physiological, psychological, economic, cultural and social elements of an Internet user which corresponds to the person's real identity" (Silva 2014, pp. 16–17). Thus, a person's cyberidentity can be understood as an extension of their identity to the Internet and any act that affects it should be illicit. Concerning the concept of OIT, different authors adopt multiple definitions. Reyns (2013) argues that identity theft is a term used to classify numerous offenses, including the fraudulent use of a person's personal information for criminal purposes without their consent. Solove (2002) adds the idea that the offender uses the stolen information to generate false information about the victim or to assume the victim's identity. For Newman and McNally (2005), the stolen information can also be used to commit other crimes, as well as to obtain financial gains. Furthermore, OIT involves the misappropriation of identity tokens such as email addresses, passwords to access online banking or webpages (Roberts et al. 2013). In the present work, OIT will be defined as *the illicit and improper use, via the Internet, of personal and financial data, which were obtained without prior consent and knowledge by the cybercriminal.*

Usually, it is possible to identify three stages in the OIT process. First, the acquisition that can be undertaken through the use of different techniques such as hacking or fraud. In the second stage, the criminal uses the stolen personal information for financial gains – the most common use – or to avoid detention by law enforcement agencies. The last stage occurs when the victim finds out that their identity has been stolen, which can happen in a short span of time or, in some cases, it can take much longer for the victim to realize (Newman and McNally 2005).

As previously mentioned, there are a set of commonly used techniques to commit OIT, which are becoming more complex and sophisticated with the development of the Internet. In fact, these techniques employ a mix of information, communication technologies and social engineering. For instance, phishing is the most common method to commit OIT (Williams 2016), involving sending deceptive e-mails that are created to appear like they are sent by legitimate businesses or organizations (Reyns and Henson 2015). Once the person receives that e-mail, the individual is redirected to a fraudulent Internet website with the single purpose of convincing them to reveal confidential information such as bank account or credit card information (Reyns and Henson 2015; Antunes and Rodrigues 2018). Along with phishing, other methods such as smishing, in which instead of sending e-mails, the offender sends text messages to the potential victim (Williams 2016), have been used. Furthermore, OIT can also be committed through the resource of pharming. This is a particular case of phishing and a more complex technique in which there is an appropriation or usurpation of the domain or URL of a legitimate webpage, which translates to the forwarding of the users to a fraudulent page in which personal information is requested. Once the fraudulent page is accessed, a virus or malicious software is installed, allowing personal or financial information to be stolen (Brody et al. 2007). Another technique used by the perpetrators is the well-known hacking, an unauthorized access, which may lead to malicious purposes such as spreading malware (Antunes and Rodrigues 2018).

5.4.2. *RAT applied to cyberspace*

The RAT developed by Cohen and Felson (1979) argued that "structural changes in routine activity patterns can influence crime rates by affecting the convergence in space and time of three minimal elements of direct-contact predatory violations: (1) motivated offenders, (2) suitable targets, and (3) the absence of capable guardians against a violation" (p. 589). This approach was developed to explain offenses and victimization that occur in the physical world, specifically in the post-World War II scenario, since authors believed that people changed their routines after that event.

With the advent of multiple forms of cybercrimes, many authors have been debating the applicability of this criminological approach to cyberspace. While Yar (2005) and Leukfeldt and Yar (2016) have been questioning the applicability of RAT to cybercrime, arguing that the convergence of victim and offender in time and space is essential, authors such as Grabosky (2001) consider cybercrimes as "new wine in old bottles", and thus defend that this approach can be applied to crimes in cyberspace. Accordingly, Eck and Clark (2003) suggest that this criminological theory can be extended to cases in which the victim and the offender are part of the geographically dispersed network, and therefore the offender is able to reach the target through the network. In these authors' point of view, the interpersonal meeting of the offender with the victim is now replaced by the network and their interaction is maintained by it, which allows the convergence of both and can lead to a possible victimization.

During the past few decades, RAT has been tested to explain various cybercrimes such as fraud, phishing, hacking, interpersonal violence or malware infection (e.g. Choi 2008; Holt and Bossler 2009; Reisig et al. 2009; Marcum et al. 2010). However, a major concern about RAT's applicability to cybercrime is the fact that few studies have empirically examined property-based crimes such as OIT victimization from a routine activities' perspective (e.g. Reyns 2013).

5.4.2.1. *Routine activities theory and cybercrime*

One of the dimensions of RAT is the proximity to offenders – online exposure – which is usually conceptualized from the victim's perspective (Reyns and Henson 2015) and has been operationalized through (i) the time spent on the Internet and/or (ii) multiple measures such as the use of online activities that can increase the risk of becoming a victim of cybercrime (e.g. online banking, social media and chat rooms) (Reyns 2013).

Previous studies found mixed results concerning the importance of online exposure to explain victimization in cyberspace. For instance, Holt and Bossler (2009) did not find a statistically significant relationship between loss of digital information by malware and online shopping, online chatting and online banking. Moreover, Leukfeldt (2014) did not observe a relationship between the activities performed online and the risk of victimization. On the other hand, Marcum et al. (2010) and Ngo and Paternoster (2011) found that the online exposure (e.g. number of hours per week that respondents engaged in instant messaging) increased the likelihood of experiencing interpersonal online victimization. Furthermore, van Wilsen (2011) observed that participation in forums and online shopping increased the likelihood of online fraud victimization and, in accordance, Reyns (2013) found that making reservations online, online social networking and purchasing behaviors were related to phishing. Specifically in what concerns OIT and RAT, a set of important results needs to be pointed out. For instance, Reyns (2013) discovered that individuals who used online banking and/or e-mail/instant messaging were about 50% more likely to become victims of OIT. Furthermore, Reyns and Henson (2015) concluded that some online activities such as online banking and purchasing contributed positively to OIT victimization.

As previously mentioned, another crucial dimension of the RAT approach is the target suitability (Cohen and Felson 1979). Generally, when applied to cyberspace, this dimension is usualy measured through the adoption of online risky activities – visits to unprotected websites, providing personal information or having personal data posted online (Reyns and Henson 2015) – since all of these activities can make the targets more attractive to the offenders (Clarke and Cornish 2000). Multiple studies have also been testing this dimension to a set of different cybercrimes. For example, Alshalan (2006) found that the more people divulge their credit or debit card numbers and disseminate their personal information, the more they are at risk of becoming cybercrime victims. Later, Ngo and Paternoster (2011) observed that clicking/opening links were activities that significantly increased the risk of acquiring a computer virus. Interestingly, Reyns (2015) showed that individuals who had their information posted online by someone else face a higher victimization risk, while those who posted accurate information themselves had a reduced victimization risk. In a more recent study, Ngo et al. (2020) found that individuals who use online banking and online travel planning had a lower risk of experiencing harassment by a non-stranger. Concerning OIT, Reyns and Henson (2015) suggested that having personal information posted online increased the victimization, and Leukfeldt and Yar (2016) found that targeted browsing, or the search for news influenced OIT.

Finally, the third element of RAT – capable guardianship – reduces the risk of victimization through voluntary actions or simply because of its presence. In the digital context, the results have been mixed, mostly because there have been different measures to operationalize the guardianship dimension (Reyns and Henson 2015). Holt and Bossler (2008) found that the physical guardian (software) did not reduce online harassment. Nevertheless, Ngo and Paternoster (2011) found a positive correlation between using secure software and malware victimization and online harassment committed by strangers. Choi (2008) discovered that college students that used a capable physical guardian had a lower risk of being victimized by a computer virus. However, Bossler and Holt (2009) and Ngo and Paternoster (2011) did not find a significant relationship between malware infection and the computer skills of the individuals. Contrarily, Holt and Bossler (2013) found that a low level of computer skills and using anti-virus software were among the significant predictors of malware infection. Holt et al. (2020) found that having a secured wireless connection decreased the risk of that crime.

Finally, considering the OIT victimization, Reyns and Henson (2015) found out that none of the items included in the capable guardian operationalization were relevant to reduce victimization. In turn, Williams (2016) tested if OIT was negatively associated with three types of guardianship: active personal, avoidance personal and passive physical guardianship. The author concluded that passive physical guardianship (e.g. using anti-virus) was effective in reducing OIT. Another interesting result was the positive association between active personal guardianship and victimization, explained by the post-victimization security reactions.

5.4.3. Individual variables and OIT victimization

Regarding the relationship between individual variables (e.g. sociodemographic characteristics) and online victimization (including OIT) mixed results have been found. For instance, Leukfeldt (2014) or van Wilsem (2013) did not find evidence that suggested the importance of individual characteristics in the likelihood of being victimized by OIT. Nevertheless, Alshalan (2006) found that more males were victims of computer viruses and cybercrime than women. Accordingly, Holt and Turner (2012) observed that males were at greater risk of OIT victimization, which is consistent with the results presented by Reyns (2013). Contrarily, Anderson (2006) found that the estimated risk of experiencing some form of OIT was 20% greater for women than for men. Concerning age, Reyns (2013) found that older adults presented a higher risk of OIT victimization, and, on the other hand, Williams (2016) and Harrell (2015) observed that younger and middle-aged adults reported more OIT victimization. In turn, Ngo and Paternoster (2011) found that each additional year in age decreased the odds of obtaining a computer virus by 2%. Recently, Bunes et al. (2020) found that individuals between the ages of 39 and 73 were at the highest risk of most types of identity theft. Moreover, these authors discovered that higher educational accomplishment was related with a higher risk of existing a credit card/bank account identity theft victimization. This result may be explained by the higher purchasing power and owning more Internet devices that store and transfer personal information (Bunes et al. 2020). Finally, results have been suggesting that people with a higher income are more likely to be victims of OIT (Reyns 2013; Reyns and Henson 2015).

Concerning social status, Williams (2016) found that lower and higher status citizens reported the highest levels of victimization, while those of average status reported the lowest, exhibiting a curvilinear association of socioeconomic status and victimization.

5.5. Fear of (online) crime

Fear of crime is a serious social problem in its own right, and, for that reason, it has been vastly studied in the last few decades (Jackson 2005). Researchers have been trying to understand not only how to best operationalize fear of crime (e.g. Gray et al. 2008), but also to discover the main individual and contextual determinants of fearing crime (e.g. Hale 1996; Guedes et al. 2018). The scientific community has consistently defined fear of crime as a phenomenon that includes three main dimensions: emotional fear of crime, cognitive risk perception of victimization and adopted behaviors for security reasons (avoidance, self-defense and protection) (e.g. Madriz 1997; Gabriel and Greve 2003). Even though fear of crime has been largely studied, little is known about the fear of online crime and specifically about the fear of OIT. In this investigation, we adopted Hille et al.'s (2015) definition of fear of OIT that includes two dimensions: the fear of financial losses and fear of reputational damage. While the first corresponds to the fear of illegal or unethical appropriation and usage of personal and financial data by unauthorized entity with the objective of getting financial benefits, the second corresponds to the fear of misuse of illegally obtained personal data with the goal of impersonation, which can cause reputational damage to the victim.

5.5.1. *Determinants of fear of (online) crime*

Gender is considered the best predictor of fear of crime (Hale 1996). However, except for domestic violence and sexual offenses, males are the most victimized, which has lead to the study and explanation of what is called the "fear-victimization paradox" (e.g. Hale 1996; Rader et al. 2007). When focusing on the fear of online crime, several studies found that women also report higher levels of fear of cybercrime, even though mixed results were also found. For instance, while Virtanen (2017) observed that women presented higher levels of fear of cybercrimes compared to men, Yu (2014) did not find any differences between women and men for fear of OIT, fraud and computer viruses. Accordingly, Roberts et al. (2013) found that gender was not a predictor of fear of OIT, concluding that this type of victimization was mainly explained by contextual dimensions and the fear of traditional crime.

Regarding age, it is possible to observe mixed results in traditional literature of fear of crime. For instance, while authors such as Reid and Konrad (2004) found that the elderly reported more fear, Ziegler and Mitchell (2003) concluded that younger people reported more fear in an experimental study. These mixed results can also be found in the cyberspace sphere. For instance, Alshalan (2006) and Lee et al. (2019) found that older individuals reported higher levels of fear of cybercrime, suggesting that older people attribute more value to property and are more afraid of losing it. Nevertheless, Henson et al. (2013) and Virtanen (2017) observed that the younger people reported more fear than older ones. Finally, Roberts et al. (2013) discovered that age, with a positive direction, was the only predictor of fear of OIT, but it accounted for less than 1% of the unique variance of fear of OIT and related fraudulent activities.

Concerning fear of traditional crime, the negative relationship between fear and socioeconomic status is well established. However, this relationship is not so clear when it comes to cybercrime (Virtanen 2017). According to Alshalan's (2006) research, it is not possible to admit a direct relationship between fear and socioeconomic status. In contrast, Virtanen (2017) and Brands and Wilsem (2019) found that individuals with lower socioeconomic status presented higher levels of fear of cybercrime. Indeed, Virtanen (2017) found that experiencing online fraud increased fear among low social status individuals, which suggests that these individuals are more affected by property-focused victimization. On the other hand, Roberts et al. (2013) found that fear of OIT is common to all socioeconomic groups. Finally, Reisig et al. (2009) found that individuals who reported lower levels of socioeconomic status perceived higher levels of risk, which was associated with online protective behaviors to avoid the risk of becoming a victim of OIT. Similarly, and taking educational levels into consideration, Akdemir (2020) found that more educated individuals reported higher levels of fear of cybercrime compared to those who were less educated. Moreover, these individuals were more likely to adopt online security measures such as the elimination of suspicious emails. In contrast, Brands and Wilsem (2019) concluded that lower levels of fear of crime were reported by individuals with higher education. Finally, both Alshalan (2006) and Roberts et al. (2013) found no direct association between education and general fear of cybercrime.

The results concerning the relationship between victimization and fear of traditional crime have been mixed (Hale 1996). Regarding cyberspace, few studies have analyzed the relationship between victimization and fear of online crime. For instance, Alshalan (2006) found that prior victimization increased the levels of fear of cybercrime, and Randa (2013) observed that previous cyberbullying victimization heightened the fear of online crime. Accordingly, Henson et al. (2013), Virtanen (2017), Lee et al. (2019), and Brands and Wilsem (2019) suggested that previous online victimization had an impact on the fear of online crime. However, Yu (2014) did not find any significant relationship between those variables.

To conclude, it is important to take into consideration the main results concerning the relationship between technical skills and the fear of online crime. It has been hypothesized that the absence of technical skills may increase the risk judgments, since individuals perceive themselves to be more vulnerable than others who are more skilful online. While Virtanen (2017) found a negative relationship between these variables, Martins (2018) observed that individuals who were considered to have less technical skills reported higher levels of fear of OIT victimization.

5.6. The present study

The main objective of the present study was to understand the impact of the Covid-19 pandemic on (i) OIT victimization and (ii) fear and perceived risk of OIT victimization. To achieve these goals, a survey previously developed by Guedes et al. (2022) was administered to a large sample, adding two groups to the original survey. Therefore, data for the current study was obtained in two phases. The first (pre-Covid-19) was collected in 2017 through an online self-report anonymous survey, disseminated both by the University of Porto (students and staff) and social media channels (N = 831). In 2021 (post-Covid-19), the same online survey was administered to a sample of 730 individuals, following the same above-mentioned sampling strategies.

5.6.1. *Measures*

Three main dependent variables are considered in this study: OIT victimization, fear of OIT victimization and perceived risk of OIT victimization. Moreover, a set of independent variables (sociodemographic, general fear of crime and online routine activity dimensions) were measured.

Dependent variables: To measure victimization of OIT, respondents were asked to rate on a scale of 0 to more than five times "how many times someone had appropriated, via the Internet, personal and financial data without prior consent and knowledge and used them improperly" during their lifetime and in the past 12 months. Prevalence of OIT victimization in a lifetime and in the last 12 months was computed, resulting in two dichotomous variables (0 = no; 1 = yes). Moreover, indirect victimization was also measured, asking the respondents to answer yes (=1) or no (=0) to the question "do you know any family member, friend or acquaintance who has had someone appropriate, via the Internet, personal and financial data without prior consent and knowledge and used them improperly?".

In turn, fear of OIT was measured through a four-point Likert scale varying between 1 (not afraid) and 4 (very afraid), asking participants *"how fearful would you be if (1) somebody stole your personal and financial data online? (2) somebody used your personal and financial data online? (3) somebody damaged your reputation based on the illegitimate use of your personal and financial data online?"*. Regarding the risk perception of victimization related to OIT, participants were asked to rate on a scale from 1 (not likely) to 5 (very likely) the following items: *"how likely do you think it is that... " (1) somebody could steal your personal and financial data online during the next 12 months? (2) somebody could use your personal and financial data online during the next 12 months? (3) somebody could damage your reputation based on the illegitimate use of your personal and financial data online during the next 12 months?*

In addition to the above-mentioned dependent variables, a survey applied post-Covid-19 included additional questions concerning the victimization of several cybercrimes, including cyberbullying, cyberstalking, hacking, phishing, OIT and malware. Respondents were asked *"in the last 24 months, have any of the following situations happened to you?"*

Sociodemographic variables: gender (0 = male, 1 = female), age (in years), perception of socio-economic status (1 = low, 2 = average, 3 = high), level of education (1 = up to four years of education to 5 = postgraduate studies) and professional situation (e.g. student, working student and self-employed) were included as sociodemographic variables. Additionally, we included a measure of fear of crime offline to understand its relationship with fear of identity theft online. Respondents were asked on a scale varying between 1 (very safe) and 5 (very unsafe): (a) *"how safe do you feel when walking alone in your residential area after dark?"* (b) *how safe do you feel being alone in your home after dark?"*.

RAT dimensions: All the dimensions arising from the RAT were operationalized in the survey: (i) online exposure to motivated offenders, (ii) online target suitability and (iii) online capable guardianship.

Online exposure to motivated offenders was measured in two different but complementary ways. First, a single item asking *"how much time, per day, in average, do you spend online?"* was used and measured in number of hours. Furthermore, the frequency of a set of online routine activities was assessed by participants on a scale of 1 (never) to 5 (always). Adapted from Reyns (2013), participants were asked *"how often do you use the Internet for the following purposes"*: (1) online banking or managing finances, (2) e-mail or instant messaging, (3) watching television or listening to the radio, (4) reading online newspapers or news websites, (5) participating in chat rooms or other forums, (6) reading or writing blogs, (7) downloading music, films or podcasts, (8) social networking (e.g. Facebook and Myspace), (9) for work or study, or (10) buying goods or services (shopping). After performing a factorial analysis with the whole sample (time 1 and time 2 survey), the items were aggregated into three types of routines: (1) financial routines (items 1 and 10), (2) working routines (items 2, 3 and 9) and (3) leisure routines (3, 5, 6 and 7). In addition to these measures, we asked participants if they generally used home banking, Paypal, credit cards, Paysafecard, MBnet and MBWAY to do their payments. Answers varied between "yes" (coded as 1) and "no" (coded as 0).

Online target suitability was measured through a set of questions adapted from previous studies conducted by Ngo and Paternoster (2011) and Reyns (2015). Concretely, participants were asked the following question: *"in the past 12 months have you"*: (1) communicated with strangers online, (2) provided personal information to somebody online, (3) opened any unfamiliar attachments in e-mails that they have received, (4) clicked on any of the web-links in emails that they have received, (5) opened any file or attachment they received through their instant messengers, (6) clicked on a pop-up message or (7) visited risky websites. Answers were measured dichotomously (yes = 1; no = 0). After performing factorial analysis, three dimensions were created: (1) interaction with strangers (items 1 and 2), (2) opening dubious links (items 3, 4 and 5) and (3) visiting risky contents (items 6 and 7).

Online capable guardianship operationalization was also adapted from previous studies (Ngo and Paternoster 2011; Williams 2016). Therefore, 13 items were used, asking participants to rate, with two options (0 = no; 1 = yes), *"for security reasons do you..."* (1) avoid online banking, (2) avoid online shopping, (3) use only one computer, (4) use an e-mail spam filter, (5) change security settings, (6) use different passwords for different sites, (7) avoid opening emails from people you do not know, (8) visits only trusted websites, (9) install and update antivirus software, (10) install and upgrade antispyware software, (11) install and update software or hardware firewalls, (12) participate in public education workshops on cybercrime or (13) visits websites aimed at public education on cybercrime. A factorial analysis was also performed and four types of security behaviors were aggregated, namely (1) avoiding behaviors (items 1 and 2), (2) protective software/hardware (items 9, 10 and 11), (3) protective behaviors (items 4–6) and, finally, (4) information/education (items 12 and 13). Furthermore, an additional question was included in order to measure participants' perception of their computer skills, varying between 1 (low) and 3 (high).

5.6.2. Results

The sociodemographic characteristics of both samples (before and after the Covid-19 pandemic) are presented in Table 5.1.

Table 5.2 presents the descriptive results of the dependent and independent variables considered in this study. The results show that victimization of OIT in the last 12 months significantly increased when compared to the period before the Covid-19 pandemic crisis ($p = .039$). In fact, 8.5% of the individuals report to have been a victim of OIT, while 5.8% were victims of OIT in the pre-Covid-19 sample. The same tendency can be observed concerning the indirect victimization, where the prevalence of individuals who reported to know any family or friends who were victims of OIT is higher in the post-Covid-19 sample ($p = .001$).

	Covid-19			
	Before (N = 831)		After (N = 730)	
	N	%	N	%
Gender/female	549	66.1	521	71.4
Education				
Up to 12th grade	330	39.7	321	43.9
Bachelor's degree	269	32.4	246	33.7
Postgraduate	232	27.9	161	22.1
Socioeconomic status				
Low	110	13.2	99	13.6
Medium	674	81.1	594	81.4
High	47	5.7	37	5.0
	Mean ± SD	Min–Max	Mean ± SD	Min–Max
Age	27.13 ± 11.07	17–68	26.29 ± 11.76	18–75

Table 5.1. *Descriptive results of the sociodemographic variables for both samples (before and after Covid-19)*

	Covid-19				
	Before		After		
	% (N)	Mean ± SD	% (N)	Mean ± SD	p^*
Victimization					
Victim of OIT (in the last 12 months)	5.8% (48)		8.5% (62)		**.036**
Indirect victimization	37.7% (347)		47.5% (660)		**.001**
Fear/risk					
General fear of crime		2.34 ± .85		2.40 ± .72	.089
Fear of OIT		3.10 ± .81		3.13 ± .75	.447
Perceived risk of OIT		2.10 ± .77		2.11 ± .67	.718

RAT					
Online exposure					
Time spent online (hours/day)		5.31 ± 3.27		5.82 ± 3.14	**.002**
Financial routines		4.83 ± 2.12		6.10 ± 2.09	**.001**
Work routines		8.53 ± 1.40		8.68 ± 1.39	**.029**
Leisure routines		8.55 ± 2.60		9.42 ± 2.69	**.001**
Homebanking	25% (205)		34.9% (234)		**.001**
Paypal	28.3% (231)		32.6% (219)		.067
Credit card	34.5% (286)		56.5% (380)		**.001**
Paysafecard	3.2% (27)		2.1% (14)		.169
MBNet	32.9% (269)		16.1% (108)		**.001**
MBWAY	–		65.1% (437)		–
Others	–		4.2% (28)		–
Target suitability					
Interaction with strangers		.41 ± .57		.46 ± .60	.067
Open dubious links		.23 ± .62		.16 ± .51	**.006**
Visit risky content		.50 ± .67		.45 ± .63	.106
Capable guardianship					
Protective software/hardware		2.41 ± .92		1.95 ± 1.06	**.001**
Avoiding behaviors		.94 ± .88		.58 ± .77	**.001**
Information		.12 ± .39		.24 ± .57	**.001**
Protective behaviors		2.35 ± .840		2.18 ± .87	**.001**
Computer skills		1.92 ± .667		1.83 ± .66	**.011**

*p-Values (two-tailed) result from t-tests or chi-square tests.

Table 5.2. *Descriptive statistics of the main variables of the present study*

Although both the direct and indirect victimization of OIT were significantly higher in the post-Covid19 sample, no changes were observed regarding the fear of OIT victimization ($p = .447$) and general fear of crime ($p = .089$), comparing the period before and after Covid-19 pandemic. The same can be observed for the risk perception of victimization, which did not alter in the considered periods ($p = .718$).

Regarding the components of RAT, results suggest that, in general, the online exposure increased in the post-Covid-19 sample. As expected, people spent more time online ($p = .002$) and used more services for financial purposes ($p = .001$), as well as for work ($p = .029$) and leisure ($p = .001$) purposes. Concerning the target suitability, operationalized through a set of risky activities, the only change observed is in the variable "open dubious links", which has significantly diminished from the pre-Covid-19 to post-Covid-19 period ($p = .006$). Regarding online capable guardianship (measured through a set of behaviors that individuals adopt to protect themselves), even though the post-Covid-19 sample reported to adopt less avoiding and protective online behaviors, it is possible to observe that they invested more in education and formation about cybercrime ($p = .001$). Nevertheless, the results suggest that the perceptions of computer skills are lower in the period after the Covid-19 pandemic ($p = .011$). Lastly, in the post-Covid-19 sample, people reported to use home banking and credit cards for online payments, rather than MBNet. Moreover, a large part of the sample uses Mbway as a form of payment (65.1%).

Cybercrimes	N	%
Phishing	447	61.5
Hacking	142	19.8
Cyberstalking	112	15.4
Malicious software	100	13.8
Cyberbullying	96	13.2
Online consumer fraud	77	10.6
Extorsion	25	3.4
OIT for crime commitment	23	3.2
OIT for fake profile creation	21	2.9
Mbway fraud	17	2.3

Table 5.3. *Prevalence of online victimization during the Covid-19 pandemic*

Table 5.3 shows the results concerning the prevalence of victimization of a list of online crimes that occurred in the last 24 months. Therefore, it is possible to observe that a large part of the sample were victims of phishing attempts (61.5%). The second more prevalent crime was hacking (19.8%) followed by cyberstalking (15.4%), malicious software (13.8%) and cyberbullying (13.2%). The mean of the variability of victimization (i.e. the mean level of the total victimizations suffered by the sample) was 1.45, varying between 0 and 9.

Finally, our data also showed that 74% of the victimization was against financial property, such as online consumer fraud, hacking, phishing, OIT for criminal purposes or Mbway fraud, and 26% was interpersonal victimization, including cybercrimes such as cyberbullying, cyberstalking, blackmail and OIT to create fake profiles.

5.6.3. *Variables associated with online victimization and fear of identity theft*

This final section intends to analyze what factors are associated with both victimization and fear of OIT. Therefore, only the sample after Covid-19 was taken into consideration. Concerning the influence of individual variables on online victimization, results suggested that gender, perception of socioeconomic status, education and age were not associated with the likelihood of being victimized. Nevertheless, when analyzing the relationship between online victimization and the contextual variables, it is possible to observe that individuals who reported to open dubious links present a higher likelihood of being victimized ($p = .015$).

Regarding fear of OIT, results showed that women reported more fear than men ($p < .001$) and an absence of correlation between age and fear. Moreover, perception of socioeconomic status was not related to fear of OIT. Interestingly, a positive correlation between this variable and the general fear of crime was observed, meaning that individuals who reported feeling insecure when walking in their residential area after dark also tend to report higher levels of fear of OIT ($r = .123$, $p < .001$). Finally, when analyzing the contextual variables, a set of important results emerged. In fact, individuals who report higher levels of fear of OIT tend to adopt more avoiding security behaviors online. Additionally, considering the target suitability, high fear is related to talking less with strangers in the cyberspace ($r = .101$, $p < .001$). Moreover, individuals with lower computer skills perception present higher levels of fear of OIT compared to individuals with advanced computer skills

($p < .001$). Lastly, results also showed that fear of OIT is higher in individuals who usually do not use Paypal as a payment method ($p < .001$).

5.7. Conclusion

The results of the present study, developed in the Portuguese context, suggest that self-reported OIT victimization (both direct and indirect) increased during the Covid-19 pandemic. These results are in accordance with previous studies that analyzed police reports of online fraud (e.g. Buil-Gil et al. 2021; Kemp et al. 2021), but also with the conclusions presented in the multiple national and international reports on cybercrime tendencies during the Covid-19 pandemic crisis. In fact, during the Covid-19 outbreak, individuals started spending more time on the Internet, exposing and expanding the frequency and variety of activities online, such as leisure, work and financial tasks (as it was also observed in the present study). These everyday routines activities, moving from the physical to the online space, may have increased the opportunities for cybercrimes to occur. Our results also showed that not only have online routine activities increased, but differences on specific forms of payments were also observed, which may be associated with the victimization of OIT. For instance, while 34.5% of the pre-Covid sample used credit cards, 56.5% reported to use this form of payment in the post-Covid sample. Home banking has also increased substantially, from 25% to 34.9%. Previous studies have found a relationship between specific forms of payments and online victimization. For instance, Alshalan (2006) concluded that the more people divulge their credit or debit card number, the more they are at risk of becoming victims of cybercrime. Nevertheless, MbNet decreased from 32.9% to 16.1%, which may be explained by the high prevalence of users of MBWAY as a form of payment. Interestingly, in spite of only 2.3% of the individuals of our study reporting to have been a victim of MBWAY fraud, this was, during the Covid-19 pandemic, one of the most reported crimes in the Portuguese context.

The results of present study also showed that, contrary to our expectations, there was a decrease in the risky activities reported by the individuals in the post-Covid sample. In fact, individuals opened less dubious links when compared to the pre-Covid sample, which may be associated with the increase in a specific guardianship behavior adopted by individuals observed in the results, that is, the search for more education and information about cybercrime. Therefore, in future studies it will be relevant to explore the relationship between the adoption of online risky activities and guardianship behaviors (e.g. Leukfeldt and Yar 2016). Concerning the variables associated with OIT victimization, and focusing on the post-Covid-19 sample, it was possible to observe that individuals who opened more dubious links were more victimized. Yet, none of the individual dimensions were relevant to the explanation of this type of victimization.

Finally, even though the scientific community has majorly studied the tendencies of cybercrimes during and after the Covid-19 pandemic crisis (mainly through statistical reports), the literature that analyzes the impact of this event on fear and perceived risk of OIT is very scarce or non-existent. In the present study, results suggested that the levels of fear of OIT and perceived risk remained the same during the considered periods, in spite of increased prevalence of OIT victimization. When analyzing the variables associated with fear of OIT, interesting results were observed, showing that both individual and contextual dimensions are relevant to explain the fear of OIT.

5.8. References

Abbasi, A., Zhang, Z., Zimbra, D., Chen, H. (2010). Detecting fake websites: The contribution of statistical learning theory. *MIS Quarterly*, 34(3), 435–461.

Abrams, D.S. (2021). Covid and crime: An early empirical look. *Journal of Public Economics*, 194, 104344.

Akdemir, N. (2020). Examining the impact of fear of cybercrime on Internet users' behavioral adaptations, privacy calculus and security intentions. *International Journal of Eurasia Social Sciences*, 11(40), 606–648.

Algarni, A., Xu, Y., Chan, T. (2017). An empirical study on the susceptibility to social engineering in social networking sites: The case of Facebook. *European Journal of Information Systems*, 26(6), 661–687.

Alshalan, A. (2006). *Cyber-Crime Fear and Victimization: An Analysis of a National Survey*. Mississippi State University, MS.

Anderson, K.B. (2006). Who are the victims of identity theft? The effect of demographics. *Journal of Public Policy and Marketing*, 25(2), 160–171.

Antunes, M. and Rodrigues, B. (2018). *Introdução à Cibersegurança: A Internet, os Aspetos Legais e a Análise Digital Forense*. FCA, Lisbon.

Bossler, A.M. and Holt, T.J. (2009). On-line activities, guardianship, and malware infection: An examination of routine activities theory. *International Journal of Cyber Criminology*, 3(1), 400–420.

Brands, J. and van Wilsem, J. (2019). Connected and fearful? Exploring fear of online financial crime, Internet behaviour and their relationship. *European Journal of Criminology*, 18(2), 1–12.

Brody, R.G., Mulig, E., Kimball, V. (2007). Phishing, pharming and identity theft. *Academy of Accounting & Financial Studies Journal*, 11(3), 43–56.

Buil-Gil, D. and Zeng, Y. (2022). Meeting you was a fake: Investigating the increase in romance fraud during Covid-19. *Journal of Financial Crime*, 29(2), 460–475.

Buil-Gil, D., Miró-Llinares, F., Moneva, A., Kemp, S., Díaz-Castaño, N. (2021). Cybercrime and shifts in opportunities during Covid-19: A preliminary analysis in the UK. *European Societies*, 23, S47–S59.

Bunes, D., DeLiema, M., Langton, L. (2020). Risk and protective factors of identity theft victimization in the United States. *Preventive Medicine Reports*, 17, 1–8.

Choi, K. (2008). An empirical assessment of an integrated theory of computer crime victimisation. *International Journal of Cyber Criminology*, 2(1), 308–333.

Cohen, L.E. and Felson, M. (1979). Social change and crime rate trends: A routine activity approach. *American Sociological Review*, 44(4), 588–608.

Dias, V.M. (2012). A problemática da investigação do cibercrime. *Data Venia Revista Jurídica Digital*, 1(1), 63–87.

Eck, J.E. and Clarke, R.V. (2003). Classifying common police problems: A routine activity approach. *Crime Prevention Studies*, 16, 7–39.

Gabriel, U. and Greve, W. (2003). The psychology of fear of crime: Conceptual and methodological perspectives. *British Journal of Criminology*, 43(1), 600–614.

Golladay, K. and Holtfreter, K. (2016). The consequences of identity theft victimization: An examination of emotional and physical health outcomes. *Victims & Offenders*, 12(5), 741–760.

Grabosky, P. (2001). Virtual criminality: Old wine in new bottles? *Social & Legal Studies*, 10(2), 243–249.

Grabosky, P. and Smith, R. (2001). Telecommunication fraud in the digital age: The convergence of technologies. In *Crime and the Internet*, Wall, D.S. (ed.). Routledge, London.

Gray, E., Jackson, J., Farral, S. (2008). Reassessing the fear of crime. *European Journal of Criminology*, 5(3), 363–380.

Guedes, I., Domingues, S., Cardoso, C. (2018). Fear of crime, personality and trait emotions: An empirical study. *European Journal of Criminology*, 15(6), 658–679.

Guedes, I., Martins, M., Cardoso, C.S. (2022). Exploring the determinants of victimization and fear of online identity theft: An empirical study. *Security Journal*.

Hale, C. (1996). Fear of crime: A review of the literature. *International Review of Victimology*, 4, 79–150.

Harrell, E. (2015). Victims of identity theft, 2014. Bureau of Justice Statistics, NCJ 248991.

Hawdon, J., Parti, K., Dearden, T.E. (2020). Cybercrime in America amid Covid-19: The initial results from a natural experiment. *American Journal of Criminal Justice*, 45(4), 546–562.

Henson, B., Reyns, B.W., Fisher, B.S. (2013). Fear of crime *online*? Examining the effect of risk, previous victimization, and exposure on fear of *online* interpersonal victimization. *Journal of Contemporary Criminal Justice*, 29(4), 475–497.

Hille, P., Walsh, G., Cleveland, M. (2015). Consumer fear of online identity theft: Scale development and validation. *Journal of Interactive Marketing*, 30, 1–19.

Holt, T.J. and Bossler, A.M. (2009). Examining the applicability of lifestyle-routine activities theory for cybercrime victimization. *Deviant Behavior*, 30(1), 1–25.

Holt, T.J. and Bossler, A.M. (2013). Examining the relationship between routine activities and malware infection indicators. *Journal of Contemporary Criminal Justice*, 29(4), 420–436.

Holt, T.J. and Turner, M.G. (2012). Examining risks and protective factors of on-line identity theft. *Deviant Behavior*, 33(4), 308–323.

Holt, T.J., van Wilsem, J., van de Weijer, S., Leukfeldt, R. (2020). Testing an integrated self-control and routine activities framework to examine malware infection victimization. *Social Science Computer Review*, 38(2), 187–206.

Horgan, S., Collier, B., Jones, R., Shepherd, L. (2021). Re-territorialising the policing of cybercrime in the post-Covid-19 era: Towards a new vision of local democratic cyber policing. *Journal of Criminal Psychology*, 11, 222–239.

Jackson, J. (2005). Validating new measures of the fear of crime. *International Journal of Social Research Methodology*, 8(4), 297–315.

Kemp, S., Buil-Gil, D., Moneva, A., Miró-Llinares, F., Díaz-Castaño, N. (2021). Empty streets, busy internet: A time-series analysis of cybercrime and fraud trends during Covid-19. *Journal of Contemporary Criminal Justice*, 37(4), 480–501.

Lallie, H.S., Shepherd, L.A., Nurse, J.R.C., Erola, A., Epiphaniou, G., Maple, C., Bellekens, X. (2021). Cyber security in the age of Covid-19: A timeline and analysis of cyber-crime and cyber-attacks during the pandemic. *Computers & Security*, 105, 102248.

Langton, S., Dixon, A., Farrell, G. (2021). Six months in: Pandemic crime trends in England and Wales. *Crime Science*, 10(1), 1–16.

Lee, S., Choi, K.., Choi, S., Englander, E. (2019). A test of structural model for fear of crime in social networking sites. *International Journal of Cybersecurity Intelligence Cybercrime*, 2(2), 5–22.

Leukfeldt, E.R. (2014). Phishing for suitable targets in the Netherlands: Routine activity theory and phishing victimization. *Cyberpsychology, Behavior, and Social Networking*, 17(8), 551–555.

Leukfeldt, E.R. and Yar, M. (2016). Applying routine activity theory to cybercrime: A theoretical and empirical analysis. *Deviant Behavior*, 37(3), 263–280.

Madriz, E. (1997). *Nothing Bad Happens to Good Girls: Fear of Crime in Women's Lives*. University of California Press, Berkeley, CA.

Marcum, C., Higgins, G., Ricketts, M. (2010). Potential factors of online victimization of youth: An examination of adolescent online behaviors utilizing routine activity theory. *Deviant Behavior*, 31(5), 381–410.

Martins, A. (2018). Sentimento de Insegurança e Vitimação no ciberespaço: relação entre variáveis individuais e contextuais. Master's Dissertation, Universidade do Porto, Porto.

Mohler, G., Bertozzi, A.L., Carter, J., Short, M.B., Sledge, D., Tita, G.E., Uchida, C.D., Brantingham, P.J. (2020). Impact of social distancing during Covid-19 pandemic on crime in Los Angeles and Indianapolis. *Journal of Criminal Justice*, 68, 101692.

Newman, G. and McNally, M. (2005). *Identity Theft Literature Review*. National Criminal Justice Reference Service, Washington, DC.

Ngo, F. and Paternoster, R. (2011). Cybercrime victimization: An examination of individual and situational level factors. *International Journal of Cyber Criminology*, 5(1), 773–793.

Ngo, F., Piquero, A., LaPrade, J., Duong, B. (2020). Victimization in cyberspace: Is it how long we spend online, what we do online, or what we post online? *Criminal Justice Review*, 45(4), 430–451.

Payne, B.K. (2020). Criminals work from home during pandemics too: A public health approach to respond to fraud and crimes against those 50 and above. *American Journal of Criminal Justice*, 45(4), 563–577.

Procuradoria-Geral da República, Gabinete do Cibercrime (2021). Cibercrime: denúncias recebidas em 2020. Nota informativa [Online]. Available at: https://cibercrime.ministeriopublico.pt/pagina/relatorio-de-atividades-33.

Rader, N., May, D., Goodrum, S. (2007). An empirical assessment of the "threat of victimization": Considering fear of crime, perceived risk, avoidance, and defensive behaviors. *Sociological Spectrum: Mid-Shouth Sociological Association*, 27(5), 475–505.

Randa, R. (2013). The influence of the cyber-social environment on fear of victimization: Cyber bullying and school. *Security Journal*, 26, 331–348.

Reid, L.W. and Konrad, M. (2004). The gender gap in fear of crime: Assessing the interactive effects of gender and perceived risk on fear of crime. *Sociological Spectrum*, 24(4), 399–425.

Reisig, M., Pratt, T., Holtfreter, K. (2009). Perceived risk of internet theft victimization: Examining the effects of social vulnerability and financial impulsivity. *Criminal Justice and Behavior*, 36(4), 369–384.

Reyns, B.W. (2013). Online routines and identity theft victimization: Further expanding routine activity theory beyond direct-contact offenses. *Journal of Research in Crime and Delinquency*, 50(2), 216–238.

Reyns, B.W. (2015). A routine activity perspective on online victimisation: Results from the Canadian General Social Survey. *Journal of Financial Crime*, 22(4), 396–411.

Reyns, B.W. and Henson, B. (2015). The thief with a thousand faces and the victim with none: Identifying determinants for online identity theft victimization with routine activity theory. *International Journal of Offender Therapy and Comparative Criminology*, 60(10), 1119–1139.

Roberts, L.D., Indermaur, D., Spiranovic, C. (2013). Fear of cyber-identity theft and related fraudulent activity. *Psychiatry, Psychology, & Law*, 20, 315–328.

Silva, F. (2014). A usurpação da ciberidentidade. Master's Dissertation, Universidade Católica do Porto, Porto.

Solove, D.J. (2002). Identity theft, privacy, and the architecture of vulnerability. *Hastings Law Journal*, 54, 1227–1276.

Van Wilsem, J. (2011). "Bought it, but never got it" assessing risk factors for online consumer fraud victimization. *European Sociological Review*, 29(2), 168–178.

Van Wilsem, J. (2013). Hacking and harassment – Do they have something in common? Comparing risk factors for online victimization. *Journal of Contemporary Criminal Justice*, 29(4), 437–453.

Virtanen, S. (2017). Fear of cybercrime in Europe: Examining the effects of victimization and vulnerabilities. *Psychiatry, Psychology and Law*, 24(3), 323–338.

Westerman, D., Spence, P.R., Van Der Heide, B. (2014). Social media as information source: Recency of updates and credibility of information. *Journal of Computer-Mediated Communication*, 19(2), 171–183.

Williams, M. (2016). Guardians upon high: An application of routine activities theory to online identity theft in Europe at the country and individual level. *British Journal of Criminology*, 56, 21–48.

Yar, M. (2005). The novelty of "cybercrime": An assessment in light of routine activity theory. *European Journal of Criminology*, 2(4), 407–427.

Yu, S. (2014). Fear of cybercrime among college students in the United States: An exploratory study. *International Journal of Cyber Criminology*, 8(1), 36–46.

Ziegler, R. and Mitchell, D. (2003). Aging and fear of crime: An experimental approach to an apparent paradox. *Experimental Aging Research*, 29(2), 173–187.

6

A South African Perspective on Cybercrime During the Pandemic

6.1. Introduction

The World Health Organization (WHO) was notified of a new coronavirus disease in December 2019, and by the end of January 2020 it had been declared a "Public Health Emergency of International Concern" [WHO 20]. Identified as "Covid-19" in February 2020, over 2 million global cases had been reported by mid-April 2020, affecting nearly every nation [EUR 20, WHO 20].

For the purposes of this chapter, the definition of cybercrime used by Stalling and Brown [STA 18] will be followed: computers or computer networks are a tool, a target or a place of criminal activity. Given the prevalence of other "smart devices" connected to networks and the Internet, these will be considered as "computers" in the above definition. This aligns with the US Department of Justice categorization of cybercrime [DOJ 00]. Essentially a computing device or network can be subjected to malicious interference that has the potential to affect the normal functioning (computer as a target); computers and networks can be used to conduct criminal activity, and computing devices can be used as a passive storage device for illegal content [DOJ 00, STA 18].

Chapter written by Brett VAN NIEKERK, Trishana RAMLUCKAN and Anna COLLARD.

The chapter investigates cybercrime activity during the Covid-19 pandemic from a South African perspective. There is disagreement to whether South Africa saw an increase in cybercrime as reported by van der Merwe [VAN 20a], or whether there was a shift in cybercrime tactics during the pandemic as indicated by Van der Walt [VAN 20b]. The chapter provides exploratory research to investigate cybercrime trends in South Africa, and attempts to clarify the question of whether there was an increase, a shift in tactics or both. International rankings, national legislature and regulations and reports of incidents are considered to analyze the trends. A limitation experienced is the lack of formal reporting of cybercrime and cyber security incidents in South Africa. Therefore, the information required is obtained from news reports on incidents and industry reports and whitepapers illustrating trends.

The chapter is structured as follows: the remainder of the introduction presents a background to South Africa's experience of Covid-19 pandemic in section 6.1.1 and describes the methodology in section 6.1.2. The three focal aspects, the international rankings, the national legislation and regulations and the incident analysis, are presented in sections 6.2–6.4, respectively. Section 6.5 discusses and triangulates the outcomes from the analysis, and the chapter is concluded in section 6.6.

6.1.1. *Background to South Africa and the pandemic*

South Africa announced a state of disaster in March 2020 in relation to the pandemic [VAN 20a]. The national disaster regulations were strict: implementing curfews, limiting sales to only essential items and outlawing the spread of disinformation. The initial intent was to delay the increase of cases to allow the health sector to prepare for what was to come.

The first wave peaked in mid-2020 and had subsided by August. The second wave occurred from December 2020 to January 2021, the third wave largely driven by the Delta variant occurred in mid-2021. At the time of writing, South Africa was experiencing its fourth wave (December 2021 to January 2022), driven by the Omicron variant. Figure 6.1 presents the new cases of Covid-19 in South Africa from January 2020 to December 2021, and the four waves can clearly be seen.

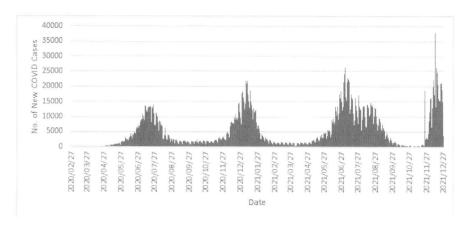

Figure 6.1. *New reported cases for Covid-19 in South Africa [WHO 21]*

Eventually, the lockdown regulations were structured into five levels, with the initial lockdown (26 March to 30 April) at level 5, reducing to level 4 until 31 May and then level 3 until 17 August. The country returned to level 3 from 29 December 2020 to 28 February 2021, and levels 3 to 4 from 16 June to 12 December 2021. Between waves, the country was on levels 2 and 1. The five alert levels had varying degrees of curfews, alcohol bans and gathering numbers. In all levels, the wearing of masks in public was mandated, with remote working recommended where possible; however, in the initial level 5 restrictions, only the essential services for organizations were able to be present on-site [SAG 21].

Another notable incident that occurred during the pandemic was the incarceration of former president Jacob Zuma, which was linked to the massive protests and looting in July 2021, particularly in the province of KwaZulu-Natal, around the major port city of Durban [BBC 21]. The relationship of these events to cyber-security incidents and disinformation will be discussed in the relevant sections.

6.1.2. *Methodology*

The objective of this chapter is to investigate cybercrime from a South African perspective during the Covid-19 pandemic. This is mostly exploratory research; however, a specific question to be answered is whether cybercrime increased during the pandemic.

Three main facets are considered: changes in cybersecurity-related international rankings of South Africa; changes in South African legislation and regulations related to cybersecurity, cybercrime, and the pandemic; and actual reports of incidents. The international rankings provide an indication of South Africa's standing, and this can infer an improving or declining cybersecurity situation. Changes in legislation can signal an impact on the cybercrime environment as certain actions can be classified as a cybercriminal, which needs to be considered during the analysis of incidents. The reported incidents give an indication to cybercrime activity and trends.

Data sources include secondary data such as international reports and previous research, and this informs both the rankings and incidents trends. Qualitative analysis legislation and related commentary, and news reports of incidents, inform both the legislation and incident aspects.

6.2. International rankings

International rankings provide a useful measure to compare national "performance" for cybersecurity. Table 6.1 lists South Africa as the origin of complaints of the FBI's Internet Crimes Complaint Centre (IC3), which is visualized in Figure 6.2. The table denotes the rank of the country within the top 20 nations from which complaints were received, and the lower the value in the ranking, the worse the reported crime rate (i.e. the 6th highest number of complaints received). Some improvements can be seen from 2014 to 2017; however, there is a noticeable decline in 2020, with nearly four times the number of complaints, with South Africa jumping from 13th most complaints in 2019 to 6th most complaints in 2020. This implies that there was either more cybercriminal activity or a greater awareness of cybercrime, resulting in more complaints. Given that there was a decrease in the number of complaints from 2014 to 2016, it can be assumed that there was already awareness and therefore an increase in cybercrime was the reasoning.

The only other African country to appear on this list is Nigeria, which ranked 16th in 2020 but has not appeared in any other reports since 2016 when it ranked 19th [IC3 17, IC3 21]. This implies that South Africa is targeted more often than any other African nations.

Year	Rank	Complaints
2012	11	–
2013	11	534
2014	11	434
2015	–	–
2016	12	337
2017	13	349
2018	13	409
2019	13	465
2020	6	1754
2021	6	1790

Table 6.1. *Annual ranking and number of complaints outside the United States to the FBI IC3 from South Africa [IC3 13, IC3 14, IC3 15, IC3 16, IC3 17, IC3 18, IC3 19, IC3 20, IC3 21, IC3 22]*

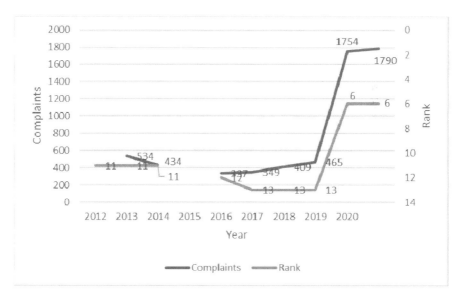

Figure 6.2. *Complaints and ranking for South Africa [IC3 13, IC3 14, IC3 15, IC3 16, IC3 17, IC3 18, IC3 19, IC3 20, IC3 21, IC3 22]*

Table 6.2 compares the ranking for cybercrime complaints reported to the FBI IC3 in 2020 and 2021, and it also compares the worldwide ranking of cumulative number of Covid cases as of December 2020 and 2021. As can be seen, of the top 10 countries for complaints received in 2020, five were in the top 10 of Covid numbers, and in 2021, six of the top 10 countries by complaints were in the top 10 for Covid numbers. For both years, Mexico and South Africa appeared in the top 20 for Covid numbers.

Complaints ranking 2020	Covid numbers rank 2020	Complaints ranking 2021	Covid numbers rank 2021
United States	1	United States	1
United Kingdom	6	United Kingdom	4
Canada	26	Canada	27
India	2	India	2
Greece	63	Australia	83
Australia	101	France	7
South Africa	17	South Africa	18
France	5	Germany	8
Germany	10	Mexico	16
Mexico	13	Brazil	3

Table 6.2. *Comparison of annual complaints ranking and cumulative Covid numbers ranking as of December 2020 and 2021 [IC3 21, IC3 22, WHO 21]*

Table 6.3 illustrates the ranking of South Africa globally and in Africa based on the International Telecommunications Union's (ITU) Cybersecurity Index. In this table, a lower value number ranking is better (i.e. "1" is top in the world). To provide more context to the South African situation, the rankings of Nigeria (chosen as it appears in the above FBI list) and Mauritius (top ranked in Africa) are also provided. As is evident, despite improving from 2017 to 2018, the ranking decreased in 2020. By comparison, Nigeria dropped in rankings between 2017 and 2018, but then improved in 2020. Mauritius has steadily dropped in the global rankings, but remains constant as the leading African nation. The rankings imply that South Africa is not making much progress compared to other nations globally and regionally; in particular, the decline in the 2020 regional ranking indicates that South Africa is lagging behind. A possible reason for Africa falling behind is that the Global North is placing greater emphasis on defense against malicious activity in cyberspace.

Year	South Africa		Nigeria		Mauritius	
	Global	Africa	Global	Africa	Global	Africa
2017	57	6	46	4	6	1
2018	56	4	57	5	14	1
2020	59	8	47	4	17	1

Table 6.3. *Annual ITU cybersecurity index rankings for South Africa [ITU 17, ITU 19, ITU 21]*

Tables 6.1 and 6.3 provide a comparison of 2020 (the pandemic) with prior years (no pandemic), and the difference is noticeable. In addition, a Bitdefender report[1] listed South Africa as 5th globally (in March 2020) and the 2nd globally (in April 2020) for countries affected by Covid-themed attacks [ARS 20]. This specifically implies a shift to specific attacks leveraging off Covid-19. A Surfshark report[2] lists South Africa as 6th for cybercrime 2021 based on the number of victims per 1 million Internet users, increasing 2% from 2020 [SUR 22].

Combining the four rankings, there is an implication of both an increase in cybercrime (based on the FBI IC3 complaints and the Surfshark report) and a shift (based on the Bitdefender report). The ITU rankings indicate that South Africa is lagging behind other nations, both globally and in Africa, in terms of the overall cybersecurity posture.

6.3. Cybercrime and related legislation

With reference to South Africa on the issues regarding cybercrimes including the fabrication and dissemination of fake news, there are a number of acts, bills and regulations which premise (implied or explicit) for the penalties associated with various cybercrimes and the spreading and creation

---

[1]. Bitdefender is a provider of Internet Security software. In April 2020, they released a report on their blog "Coronavirus-themed Threat Reports Haven't Flattened the Curve" (see [ARS 20]).
[2]. Surfshark is a provider of Internet security tools and software. In 2022, the released research shows cybercrime (see [SUR 22]).

of disinformation. The legislation includes the Electronic Communications and Transactions Act (Act 25 of 2002), South African Cybercrimes Act (Act 19 of 2020), the Protection of Personal Information Act (Act 4 of 2013), the Disaster Management Act (Regulations) and the Film Publications Board Amendment Act (Act 11 of 2019). In addition, South Africa has a National Cybersecurity Policy Framework to guide the implementation and development of cybersecurity capabilities within the country [SSA 15].

The Electronic Communications and Transmissions Act (ECT) was the initial act which introduced a number of provisions, which are now expanded upon in later acts. This includes the protection of personal information in Chapter VIII, and Cybercrime in Chapter XIII. Notably, Section 15 of the ECT Act also explicitly considers the use of "data messages" as evidence in legal proceedings. The act also regulates cryptography and authentication providers and has other clauses that relate to security [ECT 02].

The Cybercrimes Act was signed into law in 2021, with some enactments taking effect from the 21st of December 2021 [BUS 21b]. Included in the Act are the types of harmful messages that have been criminalized in South Africa. These include the incitement of damage to property or violence, the threatening of people with damage to property or violence and the unlawful containment of an intimate image [CYB 20]. However, the act does not give a specific definition for cybercrime. Therefore, fake news can be criminalized in a broader sense under the relevant legislation, and the spreading of disinformation and misinformation via digital technologies then becomes a cybercrime.

Although the Act has faced numerous criticisms including a major criticism concerning the removal of Section 17, which specifically criminalized fake news [MOY 18], it has been implied under Section 9 as Cyber forgery and uttering [CYB 20]. Furthermore, the Act creates offences for and criminalizes, among others, the disclosure of harmful data messages and disinformation spread through digital mediums such as WhatsApp. While the Cybercrimes Act took approximately 5 years and three reworks/amendments prior to enactment, it does work together with other laws and regulations. The Cybercrimes Act has a number of provisions that interact with certain aspects of the Protection of Personal Information Act (POPIA), which took almost 5 years to enact and almost 8 years to implement.

POPIA specifically regulates the methods used for the lawful processing and protection of personal information of natural and juristic persons in terms of storage and usage similar to that of the EU's General Data Protection Regulations (GDPR). In cases where any personal information becomes subject to unauthorized access or possession, POPIA addresses these issues and premises for the responsible party or parties to have taken appropriate measures to secure and safeguard the personal information first and foremost; however, when a data breach takes place, it is the duty of the responsible party to take reasonable steps to address the breach including to report the occurrence to the Information Regulator. Those responsible for any unauthorized access or possession of personal information can now also be charged with an offence under the Cybercrimes Act [BHA 21]. The POPIA has very few but specific crimes, although it did not criminalize non-compliant entities. However, one of the side effects of the Cybercrimes Act is that it ultimately criminalizes non-compliance to the POPIA, thereby making the POPIA stronger and more effective [GIL 21].

The South African Disaster Management Act Regulations are currently enforced in South Africa. While most aspects of the Regulations are intended to protect public health during the Covid-19 pandemic, Section 11(5) of the Regulations creates several content-related infringements with respect to publishing statements surrounding Covid-19. These infringements are punishable by five to six months imprisonment. More specifically, Section 11(5) criminalizes the publication, through "any medium" of information with the "intention to deceive any other person about" Covid-19, the Covid-19 infection status of any person or government measures taken in response to Covid-19 [ART 21]. Specific incidents and cases related to disinformation are discussed in section 6.4.4.

While this piece of legislation was specifically designed to counter the creation and spreading of dissemination of disinformation about the Covid-19 pandemic, this type of legislation is becoming more relevant in South Africa, although this is at a slow pace. The issue in developing such legislation is due to the creation of privacy laws such as POPIA as well as Section 16 of the South African Constitution [CON 96], which protects an individual's right to expression.

6.4. Cybersecurity incidents

As with the rest of the world, South Africa was affected by cybercrime related to the pandemic. Van der Merwe indicated that there was a 10-fold increase in affected devices in South Africa, beginning immediately after the state of disaster announcement in March 2020 [VAN 20a]. Pieterse provides a review of cyber incidents in South Africa from 2010 to 2020. While a general increase has been seen over the period, the increase from 2019 to 2020 is the largest: from 11 to 19 incidents (an increase of 8) [PIE 21]. This supports the report by van der Merwe [VAN 20a] that there has been an increase in cyber incidents.

The following sections provide an overview of notable cybercrime incidents in South Africa during the pandemic (2020–2021) and will focus on ransomware, scams, disinformation and other incidents.

6.4.1. *Ransomware*

During the pandemic, medical facilities often found themselves affected by ransomware; in South Africa, the Life Healthcare Group fell victim to this in June 2020 [BOT 20, INT 21]. Also in 2020, ransomware affected the car security and tracking company Tracker [MOY 20a] and the Office of the Chief Justice [VER 20b]; however, in the latter case data were also stolen and released [VER 20c]. A "twin attack" or "double extortion" with both ransomware and a distributed denial-of-service attack affected the social services, emergency units and banking infrastructure in October 2020 [INT 21].

In 2021, six notable ransomware attacks occurred: the Nama Khoi municipality in the Northern Cape province in May [MOY 21a]; in July, South African firms were affected by ransomware due to the global Kaseya attack [ROB 21, WHI 21], which was followed by freight logistics Transnet later in the month [GAL 21, INT 21]; the Department of Justice [NGQ 21] and debt collector Debt-IN [MOY 21c] in September; and Basil Read in December [SMI 21]. The timing of the Transnet ransomware attack was suspicious as the incident occurred the week after the political unrest in Durban due to legal proceedings against former President Zuma [MAG 21], and while possible linkages were investigated, they were considered as separate and unrelated. However, the Department of Justice incident also coincided with activity related to the case.

Sophos indicates that 24% of organizations in South Africa responded that they have experienced ransomware attacks in 2020, with an average remediation cost of 450,000 USD (ZAR 6.7 million) [SOP 21, VAN 21]. In 2019, 24% of South African organizations reported ransomware incidents, with an average remediation cost of 266,817 USD [SOP 20]. Therefore, there was an increase in ransomware incidents from 2019 to 2020; however, there was a 68.7% increase in the remediation cost. In contrast, research by Mimecast shows that ransomware prevalence increased from 45% in 2019, to 47% in 2020 and to 60% in 2021. In 2020, 53% of ransomware victims paid the ransom, although 40% of those did not get their data back; in 2021, 35% paid the ransom and 43% of those did not recover their data. The average downtime due to a ransomware incident increased from 7 days in 2020 to 11 days in 2021 [MIM 21b, MIM 22]. The average ransom paid by South African organizations in was ZAR3 261,352 [MIM 21a].

A survey by KnowBe4 and ITWeb in 2021 showed that 32% of respondents had been affected by ransomware, and 13% of those had experienced multiple incidents. In addition, a third of respondents who had suffered a ransomware attack reported data encryption, 7% experienced data encryption and exfiltration while 7% experienced all of the aforementioned as well as a distributed denial-of-service attack. While 85% of respondents indicated they were concerned about ransomware, 28% of respondents indicated their organizations were well prepared, 36% were prepared, 21% were somewhat prepared and 8% said they should be more prepared, and 7% said they were not prepared for a ransomware attack [KNO 21].

In the analysis by Pieterse [PIE 21], only one significant ransomware incident was listed prior to 2020; therefore, three major incidents in 2020 and six notable incidents in 2021 indicate a shift in cybercrime tactics targeting South Africa. Kaspersky indicated a 24% increase in ransomware in Q1 2021 to Q2 2021 [APO 21], indicating a further shift in tactics or increase in cybercrime.

Overall, there are multiple indications of an increase in ransomware during the pandemic period, as well as an increased cost of remediation and downtime due to ransomware incidents. This therefore presents a rapidly increasing risk to organizations and also highlights the fact that the discussion around cybercrime should not only focus on the quantity but also the severity.

6.4.2. Scams and fraud

With lockdown restrictions in place, the Covid-19 pandemic resulted in an increase in the utilization of online and mobile banking [SAB 21]. The South African Banking Risk Information Centre (SABRIC) reported an increase in digital banking fraud of 33% for 2020 compared to 2019 across all platforms, with an increase of 0.4% for actual monetary loss. Mobile banking was the most affected, with a 67.66% increase in incidents and a 62.1% increase in monetary loss. Online banking incidents increased by 19%; however, there was a 19 % decrease in monetary loss. The number of incidents relating to banking apps decreased by 4%, but the monetary loss increased by 14% [SAB 21].

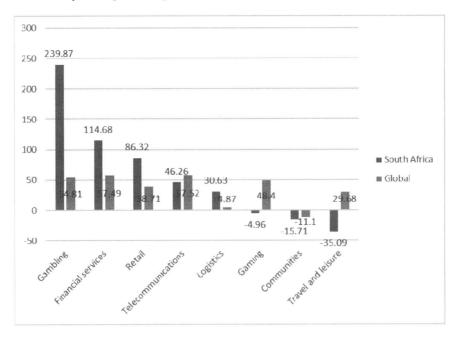

Figure 6.3. *South Africa and global changes in digital fraud (data source: [TRA 21])*

TransUnion reported a 43.62% increase in suspected digital fraudulent transaction attempts from the 2019/2020 period to the 2020/2021 period [TRA 21]. Figure 6.3 shows the changes in digital fraud across sectors after the pandemic was declared in South Africa, for both South Africa and globally. Table 6.4 illustrates the differences in the percentage change of South Africa compared to the global change; four of the eight sectors were above the global increase, with the telecommunications sector having an increase lower than the global average. The changes shown in Figure 6.3 and Table 6.4 indicate that there was an increase in cybercrime in South Africa.

Sector	South Africa vs. global increase
Logistics	528.9528%
Gambling	337.6391%
Retail	122.9915%
Financial services	99.4782%
Telecommunications	–19.5758%
Communities	41.5315%
Gaming	–110.2479%
Travel and leisure	–218.2278%

Table 6.4. *Difference in fraud from South Africa versus globally*

A range of scams and fraud were apparent in South Africa associated with the pandemic, and the most significant were false insurance, unemployment scams, fraudulent third-party sellers on online stores and phishing [BUS 21a]. Others include fake charities, shipping fraud, identity theft, account takeovers, scams related to stimulus cheques and relief funds, as well as Covid protective equipment [BUS 21a, SAC 21d]. Due to the rapid change to working from home and the rise of online meeting applications, the names of popular digital meeting and collaboration applications were used in scams or to deliver malware [STA 20a].

South Africa also saw two significant crypto-currency scams: in 2020, a Ponzi scheme stole over 500 million USD in bitcoins in what was the world's largest crypto scam at the time; then in 2021, investors lost 3.6 billion USD in a larger scam [CHE 21]. In addition to the digital fraud, accusations were raised against members of the South African government for using the pandemic as an excuse to expedite the awarding of suspect tenders for cybersecurity services [SEL 20a].

It is apparent from the data and descriptions above that South Africa experienced heightened amounts of digital fraud during the pandemic, with some sectors reporting double the number of attempted suspicious transactions. While in some instances there was a shift in tactics to use the pandemic as a lure (particularly for phishing, stimulus cheques and protective equipment), the percentage increases experienced by some sectors indicate a notable increase in cybercrime.

6.4.3. *System intrusions and data breaches*

System intrusions and data leaks continued during the pandemic. Notable incidents include the following:

– In February 2020, a third-party handling Nedbank's marketing campaigns suffered a breach which potentially affected 1.7 million customers [CIS 20].

– The chemical supplier Omnia announced it had suffered a cyberattack in March 2020, which did not affect its primary operations [BUS 20].

– In May 2020, the Unemployment Insurance Fund updated their website to incorporate the Temporary Employee/Employer Relief Scheme (TERS) implemented to assist organizations and individuals during the pandemic; this resulted in a data leak where information was exposed [VER 20a].

– It was reported in June 2020 that the Postbank's master key was printed and stolen by employees, necessitating the replacement of approximately 12 million banking cards, costing them an estimated ZAR 1 billion [STA 20b].

– In July 2020, Lombard Insurance announced it had suffered a cyber incident and that criminals may have breached customer data [MCL 20].

– Momentum Metropolitan was hacked in August 2020; however, no data were compromised [LOT 20].

– The credit firm Experian identified a breach which was made public in August 2020, where an estimated 24 million records of individuals and 793,949 businesses were exposed. In September 2020, some of the information was leaked online, and then another related leak was identified in October 2021 when information was placed on the instant messenger service Telegram [MAL 21, MOY 20b].

– The construction company Stefanutti Stocks Holdings detected a cyberattack on their systems in August 2020, resulting in the shutting down of their systems [DLA 20].

– It was reported in September 2020 that the personal information of teachers in the province of KwaZulu-Natal was leaked after they applied for "concessions" to the KwaZulu-Natal Department of Education due to their age and/or comorbidities, making them vulnerable to Covid-19 [MAG 20].

– In November 2020, Absa bank suffered a data breach due to unlawful actions by an employee [MCK 20b, OSB 20].

– The website of the African National Congress' (ANC) Youth League was hacked in November 2020 [MCK 20a].

– In December 2020, the South African Broadcasting Commission's (SABC) TV License website was compromised and was redirecting traffic to phishing websites [MCK 20c].

– PPS, and insurance group, experienced systems outages in March 2021 due to a cyberattack [BUI 21].

– The fitness group Virgin Active South Africa reported a security incident in April 2021 [MCL 21].

– A steel supplier, Macsteel, reported an unspecified cyberattack at the end of July 2021, at the same time that Transnet was affected by a ransomware incident [KHU 21].

– The South Africa National Space Agency experienced a breach of personal information in September 2021 due to an unsecured FTP connection [MOY 21b].

– The retail store Pick 'n Pay was reported to have exposed data for customers who used its delivery service in December 2021 [PRI 21]; however, they denied this in a statement, indicating that the delivery service experienced a glitch where the delivery notification link could be accessed if the correct sequence of numbers was known [MAS 21].

Despite the impending commencement of the national privacy legislation, a number of data breaches resulted from poor practice, such as having an unsecured FTP connection and mishandling of encryption keys. A concerning aspect is that organizations appear to still be unprepared even after the commencement of the privacy legislation. This indicates that there are still underlying issues making South African organizations easy targets for cybercriminals. In addition, with the commencement of the privacy and cybercrime legislation, there is greater public awareness and attention on cybersecurity issues; therefore, it is more likely that the media will report on cybersecurity incidents. This is a double-edged sword in that it encourages organizations to be more responsible in order to protect their reputation, but also highlights the weaknesses to malicious actors who may then turn their attention to those who appear to be easy targets.

6.4.4. *Disinformation and malicious communications*

With the South African legislation being amended or new legislation being introduced to cater for the digital age, disinformation and malicious communications have been outlawed. In order to mitigate pandemic related disinformation, lockdown regulations criminalized the spreading of such content [HOD 20]. In some cases, the owners of WhatsApp groups could be liable for disinformation in their group even if someone else posted it [DEV 20]. In addition, it was reported that South African legislation aimed to pressure social media organizations to take down posts containing fake news or other disinformation immediately after it was detected [DEW 20].

During the pandemic, a number of instances of disinformation or "fake news" were reported in South Africa. It is estimated that approximately 20,000 South Africans are actively engaging in anti-vaccination Facebook groups [GOO 21]. Some examples of disinformation and conspiracy theories include the following:

– altered images seemingly showing a news announcement on TV regarding the cessation of allowance payments for university students by the nation funding body [SAM 20];

– the ban on alcohol sales due to lockdown regulations being lifted early [CIL 20];

– claims that vaccines are evil through the use of religious arguments [SAC 21a];

– xenophobic claims against Indian people due to the variants emerging from India [SAC 21a];

– inadvisable and/or unapproved alternative treatments and/or medication for Covid-19 [SAC 21b, SAC 21d, SAC 21f];

– reports that Covid does not exist and/or it is a scam or hoax [SAC 21b, SAC 21d];

– Covid numbers being exaggerated [SAC 21b];

– reports of ineffectiveness and/or dangers of the vaccines [SAC 21b];

– reports that microchips or tracking devices are inside the vaccines [SAC 21b, SAC 21c];

– reports that Africans are test subjects for the vaccines [SAC 21b, SAC 21c];

– reports that the vaccines are designed to control the population in Africa [SAC 21b, SAC 21c];

– celebrities blaming deaths on the vaccine [SAC 21c];

– fake reports of new variants [SAC 21e];

– reports of infertility being caused by the vaccines [SAC 21e, SAC 21f];

– intentional underreporting of vaccine-related deaths and side effects [SAC 21f];

– information concerning how it is better to gain natural immunity rather than immunity from the vaccines [SAC 21f].

An example of these conspiracy theories are the claims that a charity tennis match in South Africa, featuring Trevor Noah and Bill Gates, was a cover-up for celebrities to arrange for the testing of vaccines in Africa. The conspiracy gained more legitimacy when a small political party formally put questions to President Ramaphosa regarding the claims, which was followed by media coverage [LER 20]. Further to the disinformation described above, social media was a significant factor during the riots and looting in July 2021 where it was used as a tool for incitement; additionally, videos unrelated to the riots were being distributed and misrepresented to inflame the situation and further divide communities [BAT 21, NET 21, PIL 21].

In addition to the legislative measures to deter disinformation, efforts to counter disinformation and malicious communications include a portal "Real 411" to enable reporting and complaints [MMA 22a], and the SA Coronavirus portal hosted weekly reports starting in May 2021 [SAC 21g]. Volunteers are also conducting fact checking and attempting to redirect people toward legitimate sources of information [GOO 21].

Table 6.5 and Figure 6.4 illustrate confirmed disinformation/misinformation cases investigated by Real 411. As is evident, disinformation peaked in Q2 of 2020 and then again in Q3 of 2021. The first peak can be explained by the ongoing pandemic and lockdown, and the second peak corresponds to the third wave. In addition, it was in between the riots and local government elections in November 2021; therefore, increased political activity could also have contributed. Figure 6.4 shows that the peaks of the disinformation tended to slightly precede the peak of waves, except in the third wave (July 2021), but this could be due to the protests and increased political activity. In terms of the platforms, Twitter and WhatsApp featured predominately, followed by Facebook.

Platform	2020				2021				Total
	Q1	Q2	Q3	Q4	Q1	Q2	Q3	Q4	
WhatsApp	17	63	26	10	22	33	33	38	242
Facebook	3	54	14	19	12	6	37	19	164
Twitter	4	33	16	60	19	28	82	40	282
YouTube		10	2	3		2	2		19
Instagram			2		1	1		2	6
Other websites	1	6	6	3	3	9	13	11	52
Political Ad					1				1
Other		9	2		3	3	6	5	28
Total	25	175	68	95	60	82	174	115	794

Table 6.5. *Disinformation/misinformation in South Africa (data source: [MMA 22b])*

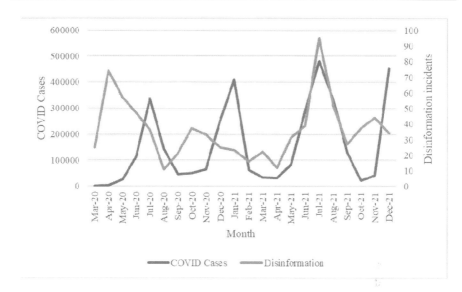

Figure 6.4. *Disinformation and Covid cases (data sources: [MMA 22b, WHO 21])*

By 24 April 2020, eight people had been arrested for spreading disinformation during the pandemic [TRU 20]. Four notable cases of legal or regulatory action regarding disinformation around the Covid pandemic include three arrests as well as action against a broadcaster by the broadcasting regulator. In April 2020, a man from Cape Town was arrested for allegedly spreading fake news on social media that the swab tests were contaminated, but the case was eventually struck from the roll due to delays and the state requesting a postponement [HYM 21]. This indicates that having legislation in place against disinformation can be ineffective if there is inadequate capacity to effectively prosecute alleged offenders. Another man was arrested for mocking the efforts to contain the virus in March 2020 [TRU 20], and a self-proclaimed prophet was arrested for circulating a video of himself denouncing the lockdown measures in April 2020 [MUK 20]. In September, the Broadcasting Complaints Commission of South Africa (BCCSA) held a hearing against the channels eNCA and eTV for airing an interview with an individual who is a Covid-denier without providing adequate opposing views. The BCCSA instructed the channels to broadcast an apology and pay a fine [BCC 20].

The legislation against disinformation will not yet lead to an increase in cybercrime figures as the reporting does not generally include disinformation, which is dealt with separately. This also raises an important question of whether disinformation should be reported as part of cybercrimes or it should be treated as a different issue altogether. Often the techniques for detecting disinformation are similar for detecting cybercrime, and there is a view that disinformation is a cybersecurity concern [IGF 21]. This discussion is outside the scope of this chapter; however, it is included due to the increasing prevalence experienced since the pandemic.

6.4.5. *Other*

In addition to the incidents described above, other cybercrime-related incidents occurred, which are not clearly categorized or have bearing on the discussion but do not fall into the period under consideration. The relevance of each case will be discussed along with the case.

A man was found guilty for hacking the emails between the victim and the broker in 2017, and sending a fraudulent email impersonating the victim's broker in order to gain authorization to transfer R350,000, enabling him to steal it [PIJ 21]. While the actual incident occurred in 2017, this illustrates the lengthy time for legal proceedings to conclude and possible delays with incidents becoming publicly known in South Africa; in this case, the pandemic may have contributed to delays. In addition, the reporting of the outcome of the incident occurred during the pandemic.

Concerns were raised around e-hailing services, where passengers have been attacked by fake drivers impersonating the service the passenger was using. This implies a degree of vulnerability in the services if imposters could get the details of the passenger [CHE 20]. There was insufficient information regarding how the incidents occurred to be able to classify it; however, it is clear that digital devices were used to commit a crime and therefore it can be classified as cybercrime.

Data from two of South Africa's major mobile service providers, provided to wireless application service providers, were linked to the assassination of a section commander in the anti-gang unit in Cape Town. Specifically, location-based data were being illegally used, and in this instance the assassins were able to track their target. This incident raised an issue of mobile networks not adequately ensuring that there was no abuse of data by the wireless application service providers; however, as a result of this incident many wireless application service providers' contracts were terminated [STA 20c]. In this instance, access to data was used to commit a crime, and therefore is related to cybercrime.

With the rapid change to remote working, the use of video conference services increased drastically, with Zoom being a popular platform. "Zoombombing" became a term where malicious actors were able to gain access to a live video conference on Zoom, and disrupt meetings. A notable incident occurred when a meeting of the South African government was zoombombed, and the malicious actors racially abused the Speaker of the National Assembly [SEL 20b]. Illegitimate access to meetings was obtained due to insecure meeting configurations, and this access was used to commit abuse, thus relating it to cybercrime.

Remote working raised the risk of cybercrime to individuals, as they did not necessarily have corporate security measures to protect them. According to the *KnowBe4 African Cybersecurity & Awareness Survey* conducted in December 2021 across eight African countries, 23% of the South African respondents said they were affected by cybercrime while working from home. Of those who were affected, 35% were exposed to a phishing attack, 17% had one of their accounts hacked, 9% fell for investment scams, 9% for ransomware, 9% for other malware and 9% for social engineering over the phone (vishing). Other scams experienced were online fraud (such as ordering something without receiving it) and theft of crypto assets (via a phished or hacked crypto wallet account) [COL 22].

The McAfee Covid-19 Threat Dashboard shows South Africa as 2nd for Covid-related malicious file detections (666,632) between February 2021 and March 2022, with the United States detecting the most with 871,608 detections [MCA 22]. Kaspersky indicated a 14% increase in crypto-mining malware in South Africa from Q1 2021 to Q2 2021 [APO 21]. These statistics imply that South Africa was a major target for cybercrime.

6.5. Discussion

Various aspects indicate that there was a definite increase in cybercrime affecting South Africa during the pandemic, such as the sudden worsening of the international rankings in 2020, with some sectors showing significant increases in cybercrime (in some cases well over the global average), while the sectors that saw a decrease only did so by a small percentage. South Africa was hard hit by Covid-themed attacks, moving from 5th to 2nd between March and April 2020, and placing 2nd overall for the number of Covid-related malicious files detected. This implies that there was both an increase in cybercrime and a shift in tactics.

Considering disinformation as a cybercrime will inflate the figures, as this categorization is fairly recent and has not been included in previous studies. However, it is still important to note that malicious actors took advantage of the situation to cause further confusion and panic. The rapid change to online and remote working saw a rise in new types of nuisance attacks, namely zoombombing, which was not prevalent before.

The ITU Global Cybersecurity Index (GCI) defines the ability of a Computer Emergency Response Team (CERT) or Computer Security Incident Response Team (CSIRT) to be accepted as a full member at the Forum for Incidence Response Teams (FIRST) as a measure of success. This type of membership serves as confirmation that the CERT/CSIRT meets the level of maturity required and verified by the global CSIRT community to deliver its services in a trustworthy and effective manner [ITU 21]. Currently, South Africa has five CERTs that are FIRST members: The National Cybersecurity Hub, the government's Electronic Communications Security – Computer Security Incident Response Team, the South African National Research Network Computer Security Incident Response Team, the Standard Bank Group's CSIRT and University of Cape Town Computer Security Incident Response Team [FIR 21]. While South Africa has more CERT/CSIRTs than some other nations, the NCPF's objective of having sector-based CERT/CSIRTs [SSA 15] has not been realized given that only 3 sectors (government, banking and academia) have such a capability. Given the disruption that the ransomware incident at the main freight organization caused, it is imperative that CERTs are established for other critical infrastructure sectors.

While South Africa enacted new legislation to address cybercrime, and the Protection of Personal Information Act became enforceable in 2021, these moves have not come soon enough to mitigate the apparent surge of cybercrime during the pandemic. There is still a need to build capacity for the law enforcement sector in order to investigate and continue with legal proceedings, as well as expanding South Africa's incident response capabilities; this will take time while the cyber criminals are already active. The delays in implementing legislation (compared to other nations) combined with the limited incident response resources in the country may have made South Africa an attractive target, explaining the worsening in the international rankings and the reports showing increases in cybercrime.

6.6. Conclusion

South Africa officially declared a State of Disaster in March 2020 in order to mitigate the rapid increase in Covid-19 infections. However, malicious actors were able to take advantage of the situation globally, targeting countries as they began seeing a noticeable increase in infections. South Africa was no different, and a number of Covid-themed attacks began to emerge. This in itself does not indicate an increase in cybercrime, but rather an opportunity for a new lure for cybercriminals to entrap their victims. However, overwhelming trends illustrated an increase in cybercrime with South Africa ranking in the top affected countries. Therefore, it has to be concluded that while there were shifts in cybercrime tactics, there was also a general increase in cybercrime during the pandemic. However, from a positive perspective, the South African legislative environment strengthened with the enactment of the Cybercrimes Act and the Protection of Personal Information Act becoming enforceable, both in 2021. However, these may have come too late, and the delays in introducing legislation could have contributed to South Africa being targeted during the pandemic. More private–public partnerships are necessary to build and increase capacity in both law enforcement and national cyber security incident response, cyber defenders and public awareness to better equip South African organizations and citizens from present and future cyber threats.

6.7. References

[APO 21] APO GROUP, Cyberattacks in Africa comparable to other parts of the globe, says Kaspersky, CNBC Africa, available at: https://www.cnbcafrica.com/2021/cyberattacks-in-africa-comparable-to-other-parts-of-the-globe-says-kaspersky/, 2021.

[ARS 20] ARSENE L., Coronavirus-themed threat reports haven't flattened the curve, Bitdefender, available at: https://labs.bitdefender.com/2020/04/coronavirus-themed-threat-reports-havent-flattened-the-curve/, 2020.

[ART 21] ARTICLE19, South Africa: Prohibitions of false Covid-19 information must be amended, Article 19, available at: https://www.article19.org/resources/prohibitions-of-false-covid-information-must-be-amended/, 2021.

[BAT 21] BATEMAN B., Fake news fuels social media hysteria, eNCA, available at: https://www.enca.com/news/crime-sa-fake-news-fuels-social-media-hysteria, 2021.

[BBC 21] BRITISH BROADCASTING CORPORATION, South Africa Zuma riots: Looting and unrest leaves 72 dead, BBC, available at: https://www.bbc.com/news/world-africa-57818215, 2021.

[BCC 20] BROADCASTING COMPLAINTS COMMISSION OF SOUTH AFRICA, Media monitoring Africa vs. eNCA Channel 403, Case number 09/2020, BCCSA, available at: https://www.bccsa.co.za/wp-content/uploads/2020/11/case-number-09-2020.pdf, 2020.

[BHA 21] BHAGATTJEE P., GOVUZA A., WESTCOTT R., Regulating the Fourth Industrial Revolution – South Africa's Cybercrimes Bill is signed into law, Cliffe Dekker Hofmeyer, available at: https://www.cliffedekkerhofmeyr.com/en/news/publications/2021/TMT/technology-media-and-telecommunications-sector-newsletter-9-june-regulating-the-fourth-industrial-revolution-south-africas-cybercrimes-bill-is-signed-into-law.html, 2021.

[BOT 20] BOTTOMLEY E., SA hit as hackers target hospitals during Covid-19 crisis – Here's what Life may be facing, Business Insider South Africa, available at: https://www.businessinsider.co.za/life-hospitals-hit-by-cyberattack-2020-6, 2020.

[BUI 21] BUSINESS INSIDER, PPS hit by cyber attack, Business Insider South Africa, available at: https://www.businessinsider.co.za/pps-hit-by-cyber-attack-report-2021-3, 2021.

[BUS 20] STAFF WRITER, JSE company Omnia hit by cyber attack, BusinessTech, available at: https://businesstech.co.za/news/it-services/381927/jse-company-omnia-hit-by-cyber-attack/, 2020.

[BUS 21a] STAFF WRITER, These are the top Covid-19 digital scams in South Africa to be aware of, BusinessTech, available at: https://businesstech.co.za/news/software/507346/these-are-the-top-covid-19-digital-scams-in-south-africa-to-be-aware-of/, 2021.

[BUS 21b] STAFF WRITER, Warning: Sending these WhatsApp messages in South Africa could now land you in jail, BusinessTech, available at: https://businesstech.co.za/news/internet/542766/warning-sending-these-whatsapp-messages-in-south-africa-could-now-land-you-in-jail/, 2021.

[CHE 20] CHECKPOINT, E-hailing hell, ENCA, available at: https://enca.com/shows/checkpoint-e-hailing-hell-06-october-2020, 2020.

[CHE 21] CHELIN R., Africa: New playground for crypto scams and money laundering, Institute for Security Studies, available at: https://issafrica.org/iss-today/africa-new-playground-for-crypto-scams-and-money-laundering, 2021.

[CIL 20] CILLIERS C., SABC warns that tweet about alcohol being sold from Monday is fake, The Citizen, available at: https://citizen.co.za/news/south-africa/social-media/2268341/sabc-warns-that-tweet-about-alcohol-being-sold-from-monday-is-fake/, 2020.

[CIS 20] CISOMAG, Nedbank's third-party data breach impacts 1.7 million customers in South Africa, EC-Council, available at: https://cisomag.eccouncil.org/nedbanks-third-party-data-breach-impacts-1-7-million-customers-in-south-africa/, 2020.

[COL 22] COLLARD A., Cyber Security Culture in SA, ITWeb [webinar], 2022.

[CON 96] CONSTITUTION OF SOUTH AFRICA, Act no. 108 of 1996, Government Gazette, Republic of South Africa, 1996.

[CYB 20] CYBERCRIMES ACT NO. 19 of 2020, Government Gazette, Republic of South Africa, 2020.

[DEV 20] DE VILLIERS J., WhatsApp group admins may be criminally liable for fake news in SA – Under these conditions, Business Insider, available at: https://www.businessinsider.co.za/whatsapp-group-admins-may-be-criminally-liable-for-fake-news-in-sa-under-these-conditions-2020-4, 2020.

[DEW 20] DE WET P., SA expects WhatsApp to "immediately" remove fake coronavirus news under new rules, Business Insider, available at: https://www.businessinsider.co.za/ott-fake-news-rules-against-coronavirus-for-whatsapp-2020-3, 2020.

[DLA 20] DLAMINI S., JSE-listed Stefanutti Stocks hit by cyberattack, engages experts to investigate, Independent Online, available at: https://www.iol.co.za/business-report/companies/jse-listed-stefanutti-stocks-hit-by-cyberattack-engages-experts-to-investigate-5e72ce33-9392-43f1-844b-0747164e98ad, 2020.

[DOJ 00] US DEPARTMENT OF JUSTICE, The electronic frontier: The challenge of unlawful conduct involving the use of the internet, US DOJ, available at: https://www.hsdl.org/?view&did=3029, 2000.

[ECT 02] ELECTRONIC COMMUNICATIONS AND TRANSMISSIONS ACT NO. 25 of 2002, Government Gazette, Republic of South Africa, 2002.

[EUR 20] EUROPEAN CENTRE FOR DISEASE PREVENTION AND CONTROL, Situation update worldwide, as of 17 April 2020, ECDC, available at: https://www.ecdc.europa.eu/en/geographical-distribution-2019-ncov-cases, 2020.

[FIR 21] FORUM FOR INCIDENCE RESPONSE AND SECURITY TEAMS, FIRST Members around the world, FIRST, available at: https://www.first.org/members/map, 2021.

[GAL 21] GALLAGHER R., BURKHARDT P., "Death Kitty" ransomware linked to South African port attack, Bloomberg, available at: https://www.bloomberg.com/news/articles/2021-07-29/-death-kitty-ransomware-linked-to-attack-on-south-african-ports, 2021.

[GIL 21] GILES J., The practical impact of the Cybercrimes Act on you, Michalsons, available at: https://www.michalsons.com/blog/the-practical-impact-of-the-cyber-bill-on-you/25300, 2021.

[HOD 20] HODGSON T.F., FARISE K., MAVEDZENGE J., Southern Africa has cracked down on fake news, but may have gone too far, Mail & Guardian, available at: https://mg.co.za/analysis/2020-04-05-southern-africa-has-cracked-down-on-fake-news-but-may-have-gone-too-far/, 2020.

[HYM 21] HYMAN A., Covid-19 "fake news" man walks free as magistrate slams prosecution, TimesLive, available at: https://www.timeslive.co.za/news/south-africa/2021-12-02-covid-19-fake-news-man-walks-free-as-magistrate-slams-prosecution/, 2021.

[IC3 13] INTERNET CRIME COMPLAINT CENTER, 2012 Internet Crime Report, Federal Bureau of Investigation, available at: https://www.ic3.gov/Home/AnnualReports, 2013.

[IC3 14] INTERNET CRIME COMPLAINT CENTER, 2013 Internet Crime Report, Federal Bureau of Investigation, available at: https://www.ic3.gov/Home/AnnualReports, 2014.

[IC3 15] INTERNET CRIME COMPLAINT CENTER, 2014 Internet Crime Report, Federal Bureau of Investigation, available at: https://www.ic3.gov/Home/AnnualReports, 2015.

[IC3 16] INTERNET CRIME COMPLAINT CENTER, 2015 Internet Crime Report, Federal Bureau of Investigation, available at: https://www.ic3.gov/Home/AnnualReports, 2016.

[IC3 17] INTERNET CRIME COMPLAINT CENTER, 2016 Internet Crime Report, Federal Bureau of Investigation, available at: https://www.ic3.gov/Home/AnnualReports, 2017.

[IC3 18] INTERNET CRIME COMPLAINT CENTER, 2017 Internet Crime Report, Federal Bureau of Investigation, available at: https://www.ic3.gov/Home/AnnualReports, 2018.

[IC3 19] INTERNET CRIME COMPLAINT CENTER, 2018 Internet Crime Report, Federal Bureau of Investigation, available at: https://www.ic3.gov/Home/AnnualReports, 2019.

[IC3 20] INTERNET CRIME COMPLAINT CENTER, 2019 Internet Crime Report, Federal Bureau of Investigation, available at: https://www.ic3.gov/Home/AnnualReports, 2020.

[IC3 21] INTERNET CRIME COMPLAINT CENTER, 2020 Internet Crime Report, Federal Bureau of Investigation, available at: https://www.ic3.gov/Home/AnnualReports, 2021.

[IC3 22] INTERNET CRIME COMPLAINT CENTER, 2021 Internet Crime Report, Federal Bureau of Investigation, available at: https://www.ic3.gov/Home/AnnualReports, 2022.

[IGF 21] INTERNET GOVERNANCE FORUM, Fighting disinformation as a cybersecurity challenge, IGF, WS #279, 7 December 2021.

[INT 21] INTERPOL, African Cyberthreat Assessment Report, Interpol, available at: https://www.interpol.int/content/download/16759/file/AfricanCyberthreatAssessment_ENGLISH.pdf, 2021.

[ITU 17] INTERNATIONAL TELECOMMUNICATION UNION, Global Cybersecurity Index 2017, ITU, available at: https://www.itu.int/dms_pub/itu-d/opb/str/D-STR-GCI.01-2017-PDF-E.pdf, 2017.

[ITU 19] INTERNATIONAL TELECOMMUNICATION UNION, Global Cybersecurity Index 2018, ITU, available at: https://www.itu.int/dms_pub/itu-d/opb/str/D-STR-GCI.01-2018-PDF-E.pdf, 2019.

[ITU 21] INTERNATIONAL TELECOMMUNICATION UNION, Global Cybersecurity Index 2020, ITU, available at: https://www.itu.int/myitu/-/media/Publications/2020-Publications/Global-Cybersecurity-Index-2020.pdf, 2021.

[KHU 21] KHUMALO S., SA steel supplier hit by cyberattack around the same time as Transnet, Fin24, available at: https://www.news24.com/fin24/companies/sa-steel-supplier-hit-by-cyberattack-around-the-same-time-as-transnet-20210804, 2021.

[KNO 21] KNOWBE4 AND ITWEB, Survey: Is SA a prime ransomware target?, ITWeb, available at: https://www.itweb.co.za/surveys/5rW1xLv5GJMRk6m3/about-results, 2021.

[LER 20] LE ROUX J., Wild Covid-19 rumours peddled by fringe political party, Daily Maverick, available at: https://www.dailymaverick.co.za/article/2020-04-17-wild-covid-19-rumours-peddled-by-fringe-political-party/, 2020.

[LOT 20] LOTZ B., Momentum Metropolitan hacked but no client data compromised, Hypertext, available at: https://htxt.co.za/2020/08/momentum-metropolitan-hacked-but-no-client-data-compromised/?amp=1, 2020.

[MAG 20] MAGUBANE K., Hundreds of teachers' personal information leaked online, Independent Online, available at: https://www.iol.co.za/mercury/news/hundreds-of-teachers-personal-information-leaked-online-9f11eb45-eff4-489b-9e15-102934539083, 2020.

[MAG 21] MAGUBANE K., Transnet "cyber attack": Govt investigating whether it's part of KZN unrest, News24, available at: https://www.news24.com/fin24/companies/transnet-unsure-when-system-will-be-restored-after-it-disruption-halts-key-operations-20210722, 2021.

[MAL 21] MALINGA S., Experian struggles to quell breach as data leaked again, ITWeb, available at: https://www.itweb.co.za/content/o1Jr5qx9OpbvKdWL, 2021.

[MAS 21] MASHEGO P., Pick n Pay denies customer data was exposed online despite "glitch", Fin24, available at: https://www.news24.com/fin24/companies/retail/pick-n-pay-denies-customer-data-was-exposed-online-despite-glitch-20211230, 2021.

[MCA 22] MCAFEE, Covid-19 related malicious file detections, McAfee, available at: https://www.mcafee.com/enterprise/en-us/lp/covid-19-dashboard.html, 2022.

[MCK 20a] MCKANE J., ANC Youth League website hacked, MyBroadband, available at: https://mybroadband.co.za/news/government/374940-anc-youth-league-websitehacked.html, 2020.

[MCK 20b] MCKANE J., Absa hit by data breach, MyBroadband, available at: https://mybroadband.co.za/news/security/378358-absa-hit-by-data-breach.html, 2020.

[MCK 20c] MCKANE J., SABC confirms that its website was hacked, MyBroadBand, available at: https://mybroadband.co.za/news/security/379914-sabc-confirms-that-its-website-was-hacked.html, 2020.

[MCL 20] MCLEOD D., SA insurer falls victim to cyberattack, TechCentral, available at: https://techcentral.co.za/sa-insurer-falls-victim-to-cyberattack/176913/, 2020.

[MCL 21] MCLEOD D., Virgin Active hit by "sophisticated" cyberattack, TechCentral, available at: https://techcentral.co.za/virgin-active-hit-by-sophisticated-cyberattack/169989/, 2021.

[MIM 21a] MIMECAST, Facing the reality gap: State of ransomware readiness, Mimecast, available at: https://www.mimecast.com/resources/ebooks/state-of-ransomware-readiness/, 2021.

[MIM 21b] MIMECAST, The state of email security 2021: Key findings in South Africa, Mimecast, available at: https://www.mimecast.com/globalassets/documents/infographics/soes-south-africa-infographic.pdf, 2021.

[MIM 22] MIMECAST, Confronting the new wave of cyberattacks: The state of email security 2022: Key findings in South Africa, Mimecast, available at: https://www.mimecast.com/globalassets/documents/infographics/2022/state-of-email-security-2022-report-infographic-south-africa.pdf, 2022.

[MMA 22a] MEDIA MONITORING AFRICA, Real 411: Fight disinformation together, Real411, available at: from https://www.real411.org/, 2022.

[MMA 22b] MEDIA MONITORING AFRICA, Real 411: See out latest trends, Real 411, available at: https://www.real411.org/trends, 2022.

[MOY 18] MOYO A., Fake news too hot to handle for Cyber Crimes Bill, IT in Government, ITWeb, available at: https://www.itweb.co.za/content/P3gQ2MGXQ4dqnRD1, 2018.

[MOY 20a] MOYO A., Tracker hack hints at more ransomware attacks in SA, ITWeb, available at: https://www.itweb.co.za/content/LPp6VMr4YxNvDKQ, 2020.

[MOY 20b] MOYO A., Data from Experian breach dumped on the Internet, ITWeb, available at: https://www.itweb.co.za/content/KA3WwMdDZeoMrydZ, 2020.

[MOY 21a] MOYO A., NCape municipality battles devastating ransomware attack, ITWeb, available at: https://www.itweb.co.za/content/8OKdWqDY581vbznQ, 2021.

[MOY 21b] MOYO A., SA's govt entities under attack as space agency hit by data breach, ITWeb, available at: https://www.itweb.co.za/content/6GxRKMYJy1bqb3Wj, 2021.

[MOY 21c] MOYO A., SA-based debt collector hit by massive ransomware attack, ITWeb, available at: https://www.itweb.co.za/content/Pero3qZxjPD7Qb6m, 2021.

[MUK 20] MUKWEVHO N., #LockdownSA: Police arrest "prophet" for circulating fake news, Health-E News, available at: https://health-e.org.za/2020/04/15/lockdownsa-police-arrest-prophet-for-circulating-fake-news/, 2020.

[NET 21] NETSWERA F., In light of the unrest and looting in South Africa, when does social media turn antisocial?, Daily Maverick, available at: https://www.dailymaverick.co.za/opinionista/2021-07-12-in-light-of-the-unrest-and-looting-in-south-africa-when-does-social-media-turn-antisocial/, 2021.

[NGQ 21] NGQAKAMBA S., Justice department's IT system brought down in ransomware attack, News24, available at: https://www.news24.com/news24/southafrica/news/justice-departments-it-system-brought-down-in-ransomware-attack-20210909, 2021.

[OSB 20] OSBORNE C., Absa bank embroiled in data leak, rogue employee accused of theft, ZDNet, available at: https://www.zdnet.com/article/absa-bank-embroiled-in-data-leak-rogue-employee-accused-of-theft/, 2020.

[PIE 21] PIETERSE H., "The cyber threat landscape in South Africa: A 10-year review", *African Journal of Information and Communication*, vol. 28, pp. 1–21, 2021.

[PIJ 21] PIJOOS I., Man found guilty of hacking broker's e-mail to swindle R350,000 from victim, TimesLive, available at: https://www.timeslive.co.za/news/south-africa/2021-10-15-man-found-guilty-of-hacking-brokers-e-mail-to-swindle-r350000-from-victim/, 2021.

[PIL 21] PILLAY Y., MTSHALI S., Social media was "instrumental" in July unrest expert tells SA Human Rights Commission, Independent Online, available at: https://www.iol.co.za/mercury/news/social-media-was-instrumental-in-july-unrest-expert-tells-sa-human-rights-commission-0d4c5257-027f-4a45-abbd-6644ce0a6415, 2021.

[PRI 21] PRIOR B., Pick n Pay customer data exposed online, MyBroadband, available at: https://mybroadband.co.za/news/security/428332-pick-n-pay-customer-data-exposed-online.html, 2021.

[ROB 21] ROBERTSON J., TURTON W., SA firms hit in massive ransomware attack, News24, available at: https://www.news24.com/fin24/companies/ict/sa-firms-also-hit-in-massive-ransomware-attack-20210705, 2021.

[SAB 21] SOUTH AFRICAN BANKING RISK INFORMATION CENTRE, Annual Crime Statistics 2020, 2021.

[SAC 21a] SOUTH AFRICAN CORONAVIRUS PORTAL, SA Covid-19 and vaccine social listening report 17 May 2021, Report 1, available at: https://sacoronavirus.co.za/2021/05/17/sa-covid-19-and-vaccine-social-listening-report-17-may2021-report-1/, 2021.

[SAC 21b] SOUTH AFRICAN CORONAVIRUS PORTAL, South Africa Covid-19 and vaccine social listening report 24 May 2021, Report 2, available at: https://sacoronavirus.co.za/2021/10/13/south-africa-covid-19-and-vaccine-social-listening-report- 24-may2021-report-2/, 2021.

[SAC 21c] SOUTH AFRICAN CORONAVIRUS PORTAL, South Africa Covid-19 and vaccine social listening report 07 June 2021, Report 4, available at: https://sacoronavirus.co.za/2021/06/07/sa-covid-19-and-vaccine-social-listening-report-07-june-2021-report-4/, 2021.

[SAC 21d] SOUTH AFRICAN CORONAVIRUS PORTAL, SA Covid-19 and vaccine social listening report 12 July 2021, Report 9, available at: https://sacoronavirus.co.za/2021/07/12/sa-covid-19-and-vaccine-social-listening-report-12-july-2021-report-9/, 2021.

[SAC 21e] SOUTH AFRICAN CORONAVIRUS PORTAL, South Africa Covid-19 and vaccine social listening report 30 August 2021, Report 16, available at: https://sacoronavirus.co.za/2021/08/30/south-africa-covid-19-and-vaccine-social-listening-report-30-august-2021-report-16/, 2021.

[SAC 21f] SOUTH AFRICAN CORONAVIRUS PORTAL, South Africa Covid-19 and vaccine social listening report 13 October 2021, Report 22, available at: https://sacoronavirus.co.za/2021/10/13/south-africa-covid-19-and-vaccine-social-listening-report-13-october-2021-report-22/, 2021.

[SAC 21g] SOUTH AFRICAN CORONAVIRUS PORTAL, Academic articles, available at: https://sacoronavirus.co.za/category/academic-articles/, 2021.

[SAG 21] SOUTH AFRICAN GOVERNMENT, About alert system, available at: https://www.gov.za/covid-19/about/about-alert-system, 2021.

[SAM 20] SAMUELS S., NSFAS warns against fake social media posts, CareersPortal, available at: https://www.careersportal.co.za/finance/nsfas-warns-against-fake-social-media-posts, 2020.

[SEL 20a] SELISHO K., DA wants Lesufi's head for R30m cyber security tender, The Citizen, available at: https://citizen.co.za/news/south-africa/politics/2266849/da-wants-lesufis-head-for-r30m-cyber-security-tender/, 2020.

[SEL 20b] SELISHO K., Hackers hurl racial abuse at Thandi Modise during virtual meeting, The Citizen, available at: https://citizen.co.za/news/south-africa/politics/2280325/hackers-hurl-racial-abuse-at-thandi-modise-during-virtual-meeting/, 2020.

[SMI 21] SMITH C., Basil Read SA's latest target of cyberattack, Fin24, available at: https://www.news24.com/fin24/companies/basil-read-sas-latest-target-of-cyberattack-20211221, 2021.

[SOP 20] SOPHOS, The state of ransomware 2020, Sophos, available at: https://www.sophos.com/en-us/medialibrary/gated-assets/white-papers/sophos-the-state-of-ransomware-2020-wp.pdf, 2020.

[SOP 21] SOPHOS, The state of ransomware 2021, Sophos, available at: https://secure2.sophos.com/en-us/medialibrary/pdfs/whitepaper/sophos-state-of-ransomware-2021-wp.pdf, 2021.

[SSA 15] STATE SECURITY AGENCY, National Cyber Security Policy Framework, Government Gazette No. 82, 4 December 2015.

[STA 18] STALLINGS W., Brown L., *Computer Security: Principles and Practice*, 4th edition, Pearson, New York, 2018.

[STA 20a] STAFF REPORTER, Criminals use names of popular social meeting applications to distribute cyber-threats, IOL, available at: https://www.iol.co.za/technology/software-and-internet/criminals-use-names-of-popular-social-meeting-applications-to-distribute-cyber-threats-46490802, 2020.

[STA 20b] STAFF WRITER, Postbank to replace 12 m bank cards after security breach, ITWeb, available at: https://www.itweb.co.za/content/nWJadvbekrmqbjO1, 2020.

[STA 20c] STAFF WRITER, Vodacom and MTN data linked to assassination scandal, MyBroadband, available at: https://mybroadband.co.za/news/cellular/371884-vodacom-and-mtn-data-linked-to-assassination-scandal.html, 2020.

[SUR 22] SURFSHARK, Cybercrime statistics, Surfshark, available at: https://surfshark.com/research/data-breach-impact/statistics, 2022.

[TRA 21] TRANSUNION, One year after Covid-19, TransUnion research shows digital fraud attempts in South Africa have increased exponentially, TransUnion, available at: https://newsroom.transunion.co.za/one-year-after-covid-19-transunion-research-shows-digital-fraud-attempts-in-south-africa-have-increased-exponentially/, 2021.

[TRU 20] TRUTER A., Misinformation and fake news on Covid-19 – Action will be taken!, LexisNexis, available at: https://www.lexisnexis.co.za/lexis-digest/resources/covid-19-resource-centre/practice-areas/media-law/misinformation-and-fake-news-on-covid-19-action-will-be-taken!, 2020.

[VAN 20a] VAN DER MERWE P., Unprecedented spike in cyber attacks since declaration of national disaster, TimesLive, available at: https://www.timeslive.co.za/news/south-africa/2020-03-26-unprecedented-spike-in-cyber-attacks-since-declaration-of-national-disaster/, 2020.

[VAN 20b] VAN DER WALT C., The impact of Covid 19 on cybersecurity, ITWeb Security Summit, 2020.

[VAN 21] VAN DER WALT C., COLLARD A., GRIMES R.A. et al., *Defending Against Ransomware: An Advisory by the South African Cybersecurity Hub*, The South African Cybesecurity Hub, Orange Cyberdefense and KnowBe4, 2021.

[VER 20a] VERMEULEN J., Data leak on UIF Covid-19 relief scheme website, MyBroadband, available at: https://mybroadband.co.za/news/cloud-hosting/353473-data-leak-onuif-covid-19-relief-scheme-website.html, 2020.

[VER 20b] VERMEULEN J., Ransomware group claims hack on Office of the Chief Justice, MyBroadband, available at: https://mybroadband.co.za/news/security/369503-ransomware-group-claims-hack-on-office-of-the-chief-justice.html, 2020.

[VER 20c] VERMEULEN J., Ransomware group releases data after attack on Office of the Chief Justice, MyBroadband, available at: https://mybroadband.co.za/news/security/374310-ransomware-group-releases-data-after-attack-on-office-of-the-chief-justice.html, 2020.

[WHI 21] WHITTAKER Z., Kaseya hack floods hundreds of companies with ransomware, TechCrunch, available at: https://techcrunch.com/2021/07/05/kaseya-hack-flood-ransomware/, 2021.

[WHO 20] WORLD HEALTH ORGANIZATION, Rolling updates on coronavirus disease (Covid-19), available at: https://www.who.int/emergencies/diseases/novel-coronavirus-2019/events-as-they-happen, 2020.

[WHO 21] WORLD HEALTH ORGANIZATION, WHO Coronavirus (Covid-19) Dashboard, available at: https://covid19.who.int/, 2021.

List of Authors

Bruno CALABRICH
Brazilian Federal Prosecutor
and
Faculty of Law
University of Brasilia
Brazil

Carla CARDOSO
School of Criminology
Faculty of Law
University of Porto
Portugal

Anna COLLARD
KnowBe4 Africa
Cape Town
South Africa

Joseph FITSANAKIS
Intelligence and Security Studies
Coastal Carolina University
United States

Inês GUEDES
School of Criminology
Faculty of Law
University of Porto
Portugal

Hugo LOISEAU
École de Politique Appliquée
Université de Sherbrooke
Quebec
Canada

Joana MARTINS
School of Criminology
Faculty of Law
University of Porto
Portugal

Alexa MCMICHAEL
Special Security Officer
Intelligence Operations Command
Center
Coastal Carolina University
United States

Samuel MOREIRA
School of Criminology
Faculty of Law
University of Porto
Portugal

Brett VAN NIEKERK
Durban University of Technology
South Africa

Trishana RAMLUCKAN
University of KwaZulu-Natal
Durban
South Africa

Daniel VENTRE
CNRS
CESDIP Laboratory
Guyancourt
France

Alexandre VERONESE
Social and Legal Theory
Faculty of Law
University of Brasilia
Brazil

Index

A, B

accelerationist/accelerationism, 82, 83, 91–94, 97, 99–102
adequacy level, 111, 116, 117
APT32, 24
attacks, 2, 20, 23, 24, 29–34, 37, 38, 40, 81, 82, 85, 98, 100, 101
  cyberattacks, 24, 29–35, 37, 38
BRICS (Brazil, Russia, India, China and South Africa), 111, 132
Budapest Convention, 111, 126, 129–133

C

Canada, 49–52, 60, 61, 67–72
CERT (Computer Emergency Response Teams), 5, 6, 11–14, 18–21, 37, 38, 41
content, 120–124, 126, 128, 138–140
Covid-19, 49, 51, 52, 57, 63, 67, 69
criminals, 2, 24, 36, 37, 42
  cybercriminals, 1, 2, 6, 18, 23, 27, 35, 37, 38
cyber risk, 52, 53, 62, 71

cybercrime (*see also* criminals), 49, 50, 52, 53, 55–74, 81, 83–85, 100–102, 150–154, 156–163, 165, 168–171, 177–180, 182–190, 192, 196–199
cybersecurity, 52, 53, 58, 60, 62, 66, 68–71, 73, 74, 180, 182–184, 186, 190, 192, 196–198
cyberspace, 49, 50, 52, 53, 55–60, 62, 64, 69, 72, 74

D, E, F

data breach, 185, 190–192
disinformation, 178, 179, 184–186, 192–196, 198
exposure, 157, 164, 167, 168
extremist/extremism, 82–84, 86, 95–99, 101, 102
fear, 149, 150, 154, 160–164, 166, 168–171
Federal Prosecutors Office, 109, 110, 124
fraud, 54, 55, 58, 62–64, 68, 149, 151–155, 157, 160, 161, 168–170, 188–190, 197

H, I

health crisis, 50, 52
identity theft, 149, 152, 154, 155, 159, 164, 169
impact, 1, 4, 9, 37, 38, 150, 162, 171
incident, 5, 11–21, 26–31, 37
insurrection, 81, 88, 89, 100

L

law enforcement, 109, 113–115, 121, 122, 124–127, 130, 134, 140, 147
legal interoperability, 111, 115–119, 132, 141
LICRA versus Yahoo, 122, 137, 139

M, N, O, P

militant/militancy, 81–86, 88, 89, 91, 92, 94–102
North America, 51, 53, 55, 61, 71, 72
online, 81–86, 89–92, 94–100, 102, 149–155, 157–165, 167–171
pandemic, 1–4, 7, 9, 18, 19, 24, 28–30, 35–38, 49–57, 60–65, 67–74, 177–180, 183, 185–190, 192, 194–199
perception, 150, 154, 160, 163–165, 168, 169
privacy legislation/laws, 185, 192

R, S

radicalization, 81, 84–86, 92, 94–98, 100–102
ransomware, 57, 62, 65–70, 186, 187, 191, 197, 198
riot, 91
risks, 1, 2, 4, 18, 21, 35, 38, 40, 46, 150, 154, 157–163, 166, 168, 170, 171
routine activities, 150, 156, 157, 163, 164
SaferNet, 109–111, 120–124, 126
SARS-CoV-2 pandemic, 49, 50, 52, 53, 60, 61, 69, 70, 72
social media, 83, 86, 90, 93–99, 101, 102
statistics, 2–6, 9, 11, 13, 18, 30, 34, 36–38, 41, 45, 46

T, U, V

target, 150, 151, 156, 158, 164, 167–169
trends, 4, 9, 11–13, 18, 24, 28, 29, 36, 37, 40
United States, 49–54, 60–62, 65–68, 70–72
victimization, 149–152, 154, 156–163, 165, 166, 168–171
cybervictimization, 152

Other titles from

in

Science, Society and New Technologies

2022

AIT HADDOU Hassan, TOUBANOS Dimitri, VILLIEN Philippe
Ecological Transition in Education and Research

CARDON Alain
Information Organization of The Universe and Living Things: Generation of Space, Quantum and Molecular Elements, Coactive Generation of Living Organisms and Multiagent Model
(Digital Science Set – Volume 3)

CAULI Marie, FAVIER Laurence, JEANNAS Jean-Yves
Digital Dictionary

DAVERNE-BAILLY Carole, WITTORSKI Richard
Research Methodology in Education and Training: Postures, Practices and Forms
(Education Set – Volume 12)

ELAMÉ Esoh
Sustainable Intercultural Urbanism at the Service of the African City of Tomorrow
(Territory Development Set – Volume 1)

FLEURET Sébastien
A Back and Forth Between Tourism and Health: From Medical Tourism to Global Health
(Tourism and Mobility Systems Set – Volume 5)

KAMPELIS Nikos, KOLOKOTSA Denia
Smart Zero-energy Buildings and Communities for Smart Grids
(Engineering, Energy and Architecture Set – Volume 9)

2021

BARDIOT Clarisse
Performing Arts and Digital Humanities: From Traces to Data
(Traces Set – Volume 5)

BENSRHAIR Abdelaziz, BAPIN Thierry
From AI to Autonomous and Connected Vehicles: Advanced Driver-Assistance Systems (ADAS)
(Digital Science Set – Volume 2)

DOUAY Nicolas, MINJA Michael
Urban Planning for Transitions

GALINON-MÉLÉNEC Béatrice
The Trace Odyssey 1: A Journey Beyond Appearances
(Traces Set – Volume 4)

HENRY Antoine
Platform and Collective Intelligence: Digital Ecosystem of Organizations

LE LAY Stéphane, SAVIGNAC Emmanuelle, LÉNEL Pierre, FRANCES Jean
The Gamification of Society
(Research, Innovative Theories and Methods in SSH Set – Volume 2)

RADI Bouchaïb, EL HAMI Abdelkhalak
Optimizations and Programming: Linear, Non-linear, Dynamic, Stochastic and Applications with Matlab
(Digital Science Set – Volume 1)

2020

BARNOUIN Jacques
The World's Construction Mechanism: Trajectories, Imbalances and the Future of Societies
(Interdisciplinarity between Biological Sciences and Social Sciences Set – Volume 4)

ÇAĞLAR Nur, CURULLI Irene G., SIPAHIOĞLU Işıl Ruhi, MAVROMATIDIS Lazaros
Thresholds in Architectural Education (Engineering, Energy and Architecture Set – Volume 7)

DUBOIS Michel J.F.
Humans in the Making: In the Beginning was Technique
(Social Interdisciplinarity Set – Volume 4)

ETCHEVERRIA Olivier
The Restaurant, A Geographical Approach: From Invention to Gourmet Tourist Destinations
(Tourism and Mobility Systems Set – Volume 3)

GREFE GWENAËLLE, PEYRAT-GUILLARD DOMINIQUE
Shapes of Tourism Employment: HRM in the Worlds of Hotels and Air Transport (Tourism and Mobility Systems Set – Volume 4)

JEANNERET Yves
The Trace Factory
(Traces Set – Volume 3)

KATSAFADOS Petros, MAVROMATIDIS Elias, SPYROU Christos
Numerical Weather Prediction and Data Assimilation (Engineering, Energy and Architecture Set – Volume 6)

KOLOKOTSA Denia, KAMPELIS Nikos
Smart Buildings, Smart Communities and Demand Response (Engineering, Energy and Architecture Set – Volume 8)

MARTI Caroline
Cultural Mediations of Brands: Unadvertization and Quest for Authority
(Communication Approaches to Commercial Mediation Set – Volume 1)

MAVROMATIDIS Lazaros E.
Climatic Heterotopias as Spaces of Inclusion: Sew Up the Urban Fabric
(Research in Architectural Education Set – Volume 1)

MOURATIDOU Eleni
Re-presentation Policies of the Fashion Industry: Discourse, Apparatus and Power (Communication Approaches to Commercial Mediation Set – Volume 2)

SCHMITT Daniel, THÉBAULT Marine, BURCZYKOWSKI Ludovic
Image Beyond the Screen: Projection Mapping

VIOLIER Philippe, with the collaboration of TAUNAY Benjamin
The Tourist Places of the World
(Tourism and Mobility Systems Set – Volume 2)

2019

BRIANÇON Muriel
The Meaning of Otherness in Education: Stakes, Forms, Process, Thoughts and Transfers
(Education Set – Volume 3)

DESCHAMPS Jacqueline
Mediation: A Concept for Information and Communication Sciences
(Concepts to Conceive 21st Century Society Set – Volume 1)

DOUSSET Laurent, PARK Sejin, GUILLE-ESCURET Georges
Kinship, Ecology and History: Renewal of Conjunctures
(Interdisciplinarity between Biological Sciences and Social Sciences Set – Volume 3)

DUPONT Olivier
Power
(Concepts to Conceive 21st Century Society Set – Volume 2)

FERRARATO Coline
Prospective Philosophy of Software: A Simondonian Study

GUAAYBESS Tourya
The Media in Arab Countries: From Development Theories to Cooperation Policies

HAGÈGE Hélène
Education for Responsibility
(Education Set – Volume 4)

LARDELLIER Pascal
The Ritual Institution of Society
(Traces Set – Volume 2)

LARROCHE Valérie
The Dispositif
(Concepts to Conceive 21st Century Society Set – Volume 3)

LATERRASSE Jean
Transport and Town Planning: The City in Search of Sustainable Development

LENOIR Virgil Cristian
Ethically Structured Processes
(Innovation and Responsibility Set – Volume 4)

LOPEZ Fanny, PELLEGRINO Margot, COUTARD Olivier
Local Energy Autonomy: Spaces, Scales, Politics
(Urban Engineering Set – Volume 1)

METZGER Jean-Paul
Discourse: A Concept for Information and Communication Sciences
(Concepts to Conceive 21st Century Society Set – Volume 4)

MICHA Irini, VAIOU Dina
Alternative Takes to the City
(Engineering, Energy and Architecture Set – Volume 5)

PÉLISSIER Chrysta
Learner Support in Online Learning Environments

PIETTE Albert
Theoretical Anthropology or How to Observe a Human Being
(Research, Innovative Theories and Methods in SSH Set – Volume 1)

PIRIOU Jérôme
The Tourist Region: A Co-Construction of Tourism Stakeholders
(Tourism and Mobility Systems Set – Volume 1)

PUMAIN Denise
Geographical Modeling: Cities and Territories
(Modeling Methodologies in Social Sciences Set – Volume 2)

WALDECK Roger
Methods and Interdisciplinarity
(Modeling Methodologies in Social Sciences Set – Volume 1)

2018

BARTHES Angela, CHAMPOLLION Pierre, ALPE Yves
Evolutions of the Complex Relationship Between Education and Territories
(Education Set – Volume 1)

BÉRANGER Jérôme
The Algorithmic Code of Ethics: Ethics at the Bedside of the Digital Revolution
(Technological Prospects and Social Applications Set – Volume 2)

DUGUÉ Bernard
Time, Emergences and Communications
(Engineering, Energy and Architecture Set – Volume 4)

GEORGANTOPOULOU Christina G., GEORGANTOPOULOS George A.
Fluid Mechanics in Channel, Pipe and Aerodynamic Design Geometries 1
(Engineering, Energy and Architecture Set – Volume 2)

GEORGANTOPOULOU Christina G., GEORGANTOPOULOS George A.
Fluid Mechanics in Channel, Pipe and Aerodynamic Design Geometries 2
(Engineering, Energy and Architecture Set – Volume 3)

GUILLE-ESCURET Georges
Social Structures and Natural Systems: Is a Scientific Assemblage Workable?
(Social Interdisciplinarity Set – Volume 2)

LARINI Michel, BARTHES Angela
Quantitative and Statistical Data in Education: From Data Collection to Data Processing
(Education Set – Volume 2)

LELEU-MERVIEL Sylvie
Informational Tracking
(Traces Set – Volume 1)

SALGUES Bruno
Society 5.0: Industry of the Future, Technologies, Methods and Tools
(Technological Prospects and Social Applications Set – Volume 1)

TRESTINI Marc
Modeling of Next Generation Digital Learning Environments: Complex Systems Theory

2017

ANICHINI Giulia, CARRARO Flavia, GESLIN Philippe,
GUILLE-ESCURET Georges
Technicity vs Scientificity – Complementarities and Rivalries
(Interdisciplinarity between Biological Sciences and Social Sciences Set – Volume 2)

DUGUÉ Bernard
Information and the World Stage – From Philosophy to Science, the World of Forms and Communications
(Engineering, Energy and Architecture Set – Volume 1)

GESLIN Philippe
Inside Anthropotechnology – User and Culture Centered Experience
(Social Interdisciplinarity Set – Volume 1)

GORIA Stéphane
Methods and Tools for Creative Competitive Intelligence

KEMBELLEC Gérald, BROUDOUS EVELYNE
Reading and Writing Knowledge in Scientific Communities: Digital Humanities and Knowledge Construction

MAESSCHALCK Marc
Reflexive Governance for Research and Innovative Knowledge
(Responsible Research and Innovation Set - Volume 6)

PARK Sejin, GUILLE-ESCURET Georges
Sociobiology vs Socioecology: Consequences of an Unraveling Debate
(Interdisciplinarity between Biological Sciences and Social Sciences Set – Volume 1)

PELLÉ Sophie
Business, Innovation and Responsibility
(Responsible Research and Innovation Set – Volume 7)

2016

BRONNER Gérald
Belief and Misbelief Asymmetry on the Internet

EL FALLAH SEGHROUCHNI Amal, ISHIKAWA Fuyuki, HÉRAULT Laurent, TOKUDA Hideyuki
Enablers for Smart Cities

GIANNI Robert
Responsibility and Freedom
(Responsible Research and Innovation Set – Volume 2)

GRUNWALD Armin
The Hermeneutic Side of Responsible Research and Innovation
(Responsible Research and Innovation Set – Volume 5)

LAGRAÑA Fernando
E-mail and Behavioral Changes: Uses and Misuses of Electronic Communications

LENOIR Virgil Cristian
Ethical Efficiency: Responsibility and Contingency
(Responsible Research and Innovation Set – Volume 1)

MAESSCHALCK Marc
Reflexive Governance for Research and Innovative Knowledge
(Responsible Research and Innovation Set – Volume 6)

PELLÉ Sophie, REBER Bernard
From Ethical Review to Responsible Research and Innovation
(Responsible Research and Innovation Set – Volume 3)

REBER Bernard
Precautionary Principle, Pluralism and Deliberation: Sciences and Ethics
(Responsible Research and Innovation Set – Volume 4)

VENTRE Daniel
Information Warfare – 2nd edition